A
CARELESS
BUSINESS

AN INSIDER'S ACCOUNT
OF SOCIAL CARE IN THE UK

A. CAREWORKER

ISBN: 978-1-4834-3099-7 (sc)
ISBN: 978-1-4834-3098-0 (e)

Lulu Publishing Services rev. date: 06/24/2015

"It is not our business here to consider what bearing the permanent existence of…… a solid layer of savagery beneath the surface of society, and unaffected by superficial changes of religion or culture, has upon the future of humanity. The dispassionate observer, whose studies have led him to plumb its depths, can hardly regard it otherwise than as a standing menace to civilization. We seem to move on a thin crust which may at any moment be rent by the subterranean forces slumbering below. From time to time a hollow murmur underground or a sudden spirt of flame into the air tells us of what is going on beneath our feet."

James George Frazer – The Golden Bough.

FOREWORD

This was a book I never intended to write, and perhaps that is the best reason for writing anything, as to have an idea of producing some work, believing it to be of importance, consequence or significance and with a sense of producing some great "statement" can result in the most baleful and terrible of results, whereas writing something because you feel the absolute compulsion to write it, against all your better reasons and judgement, is maybe because you should and have to do it.

I say I never intended to write this book because going into care was not some enterprise by which I could see what it was like "on the inside", to learn about it or with the intension of later exposing it's seamier side; it was a career choice, something I wanted to do, more, was encouraged to do because people already working in care told me I may have an aptitude for it.

So I decided to give it try.

I had no qualifications, no experience and no idea just what it was like to actually be a carer or support worker; as the following will show that was no barrier to entry and it did not take long for my eyes to be fully opened. Yet for all I was shocked virtually from day one I had no thought of ever committing my experiences to the page because to do so would obviously be to have committed professional suicide – if you're working in a job, any job, the best way out of it is to tell as many people as possible just what it is like to actually do it, especially the pressures involved and the cynical side of it. However as time went on, and, most specifically when I went to work in the private – and for profit - care for the elderly system, I became ever more acutely aware that what I was doing, what I was asked to do, and how the system as a whole controlled and compelled such actions, was at its and foundation and basis fundamentally wrong.

With the passage of time two realisations slowly overcame me, the first was the conviction that if the public at large knew how society treated and "cared" for its elderly and infirm and physically or mentally challenged – without any laws being broken, and with the tacit consent of watchdogs and oversight bodies - then they would be appalled; appalled and ashamed that in a 21st century rich world country the last or the longest years of many old and vulnerable people were spent, or should I say endured, in a substandard environment that put financial considerations above their welfare. I was sure that if the public knew just how utterly broken, unjust and corrupt the system is, and how private profit is made from the public purse - as well as the estates of merely modestly well off old people - then they would have serious questions to ask not just of care providers but of a political class of every stripe that has allowed such a system to spring up in the first place, and then flourish almost totally unhindered and almost completely unbounded.

The second realisation was that the longer I spent working in the care system the more I was becoming desensitised to all I saw happening around me, worse than that though - I was actually becoming complicit in it. I no longer questioned why this or that thing was allowed to happen, I just followed orders, and, at times, committed some of the very actions that initially disgusted me; in other words, and in simple terms, I was becoming part of the problem.

One day it struck me just how indifferent I had become, how utterly unfeeling and disgustingly compromised I was - I was coming to hate the person I was, at least as a carer. This had not happened all at once, it had occurred over time, some of it was due to the constant drip, drip, drip of pressures to meet expectations that had nothing to do with the care of the people I was looking after but rather to meet what managers, senior care staff and other carers expected me to do - for appearances sake or their convenience; however I cannot use this excuse for everything, for I too was organizing people's care to fit with my own convenience; I realised that I occupied no moral high ground and was intoxicated more by hypocrisy than justice. Eventually I had to ask myself how I would or could account for my actions, not if I was brought to book professionally, but if I was to account for them morally – I came to the conclusion I couldn't.

The obvious answer would have been to quit and find another way to earn a crust, but at the time of writing, in the middle of a deep recession, I couldn't just walk away and into another job, neither really did I much want to because there were moments when I caught a glimpse of the carer I could be. At times - freed from the almost unremitting and huge pressure that was usually heaped upon me and those I worked with - I found I could be a good carer, that given the right situation I could be the person that professionally I always wanted to be. This left me with the thought that if the system itself were changed then perhaps care did not have to be the way it is currently condemned to be, but how would that happen?

Here I will avoid the grandiose; in writing about my experiences I was, and am, under no illusion that I can change anything, the financial imperatives are too great and the whole care apparatus – and vested interests - are too huge and complex to be undone simply by words, it would take a great act of will and money to create a care system that was worthy of the name, and neither will be forthcoming anytime soon. However I also recognise that the care system as it stands is hugely wasteful, inefficient and therefore unnecessarily expensive, it does not all come down to throwing money at the issue, there are many things that can be fixed not only with no extra cost but actually at a saving.

So why write?

There are several reasons which I will briefly mention.

The first was, as noted above, to show the public what care for the elderly and vulnerable is like from the inside, why things are as they are and how they result in situations in which relatives, friends and loved ones suffer needlessly, not in any spectacular fashion – preventable deaths and outright abuse are mercifully rare, contrary to popular opinion – but in small degrading steps, small indignities and privation of liberties in ways that are beyond the obvious.

The second reason is again noted above; it is to try to rescue the carer I would like to be from the one I felt I was becoming, that by writing down just why things have turned out as they have I can confess how I was sucked into the machine that chews well-meaning Care Assistants (CW's) to fragmented pieces that function only at a task orientated indifferent level, or spits them out disillusioned, disheartened and even dismissed.

As well as these though there are two more reasons for writing, one is to put, for the first time to my knowledge, the care worker's point of view. It was a – very black - joke I shared with my partner that I was seriously considering telling people I was released murderer rather than a Care Assistant as the former seemed to have the better reputation than the latter; this was only half and ironic jest. In nearly all the scandals that have broken in and about the care sector it is CW's that have been the focus. Some of this is undoubtedly deserved; no one who saw the footage from the Winterbourne View case could witness the premeditation and sadism in the acts perpetrated as anything other than the actions of warped and dangerous people. Other scandals have followed that have been less incendiary but, in their way, equally as damaging, and in nearly all these cases the first to be thrown under the proverbial bus were CW's while senior managers and companies have been let off the hook or made out with handsome payoffs.

Often, in situations where CW's fail in their duty of care or crack under pressure, it is the result of the structure of care that places them in impossible situations; abuse and neglect are always inexcusable but this is not the same as saying that they do not have deep rooted foundations beyond the CW's' control. Therefore I wanted to show just what it is like to be a care worker, how so many competing demands are made on you that it sometimes means that in the striving to meet all of them none of them actually get met and how the pressure of the job can lead to aberrant reactions that in isolation look terrible but are in fact a consequence of corners being cut further up the pay scale.

There too is the need for me to show that there are very many good carers who labour in very difficult circumstances and never get noticed, simply because they do their job fantastically well without fanfare. Many of these I have learnt from and have made me a better CA and, often times, a better person. To these unsung labourers I owe a debt and I want to show how it is often these least noticed CW's who are among the best.

Finally then two notes about the following –

Firstly to write a purely narrative account of my time in care is simply not an option for this CA as it could not be done without compromising the identity of those I cared for and those I worked with – it would be a betrayal of the very core of care values that I hold to be fundamental – those of dignity, respect and confidentiality; so, although possible it would be a contradiction of the aim of this work. Therefore I have chosen a middle course in the following structure, I have used my personal experiences as the foundation for exploring the ills and failures of care – a kind of jumping off point – in which I will move from the personal to the universal, this hopefully will give the reader both an insight not my own care career but, of infinitely more value, it will link this persal account to matters beyond and broader to the purely individual.

Also in the narrative I have included some situations that I did not witness first hand but the accounts of which have been passed on to me by reliable sources that have had corroboration from others and that therefore I can credibly recount.

The second note is that this is not a purely scientific, medical, social or economic study of care - although it includes references to such sources - it is not meant to be an elongated completely factual essay about care provision and delivery in the UK at present but *my thoughts and observations borne of my experience of care provision*, it is then an emotionally human rather than an academically disinterested response to my work in care; it is a moderated *cri de coeur* against the unfeeling edifice of ruthless market driven social care.

In the structure of the book I hope to have avoided two potential pitfalls that glared at me during the writing of it, the first is self-indulgence and the second indifference; I do not want this account to be overwrought with shrill prose though neither do I wish to convey anything less than a total emotional immersion and stake in the matters I cover.

Although many people have helped and advised me during the writing of this book any errors are my own, and being human and more fallible than most, the fault is all mine.

A NOTE ON TERMINOLOGY

In the care sector a proliferation of titles exists that cover much of the same tasks, functions and job descriptions of care and clinical staff. In addition those in receipt of care are also classified under varied nomenclature. For ease of reading I have condensed such titles even though those within care may object that I have used the same title for jobs that are subtly different. This issue cannot be squared so I have come down on the side of a simplified number of titles to make it easier for a person not versed in care to understand.

Care Assistants (CW's) – under this title I include support workers for those with learning or physical challenges, Health Care Assistants (HCW's) - a title most often found within the National Health Service (NHS) - as well as carers either in a residential setting or in domiciliary care (home care services).

Service Users – although I recognise the rather antiseptic and non-individualised nature of the terminology I have employed, this term however most accurately describes the situation of those in receipt of care – people who are reliant on care services. This term is not intended to diminish, degrade or disempower those it describes; it is just the simplest term for individuals in receipt of care and therefore is the most inclusive. Further it most aptly describes the attitude and outlook of care providers to those who pay (directly or indirectly) for their services.

Clinicians – this term refers to those with formal medical training such as nurses and doctors and is used to differentiate them from CW's who are non-medically trained (although it is recognised that they may undertake some tasks that normally would be associated with clinicians).

INTRODUCTION

When I Say Crisis
I Mean Crisis

Care in crisis.

I was going to call the book this; at the very least I was going to title this chapter with these words. However I realised that this phrase had been used so often that it had lost all meaning and significance. Instead of a tocsin or clarion call it has become a mantra that, rather than causing alarm, has induced only a narcotized sense of indifference. We read the words or hear the phrase, perhaps cock one ear toward the report that follows or cast a lazy eye over the article and then move on. We have become inured to the crisis in care which seems, like the poor, to have always been with us. The fact is though it is the truth, and the truth is as raw as burnt skin, and the longer time goes on, the rawer the truth; for care is now not just a ship headed for the rocks, it is a shipwreck and the only question that remains is how many survivors can be rescued.

For a long time care – or to be more accurate adult care – of which this book covers – has been in crisis. It is the Cinderella service. While we have become deeply sensitised as a society to the protection and care of vulnerable children we have become in equal measure desensitised to the failings of adult care. We assume that if adults needing social care are not being beaten, starved or dehydrated to death, chronically neglected, forced to live in their own filth or murdered then everything must be ok – right?

Wrong. Adult care is failing in its very basics – caring - because while there are a few instances of all of the above – creating spectacular headlines and causing politicians to rush about expending much hot air speechifying, commissioning volumes of heated prose by assorted worthies in weighty reports that no-one ever reads, but precious little light or relief – the real failings go on unaddressed.

These problems are chronic failings which I ran into every day in the course of my time as a Care Assistant (CA). They can ultimately be summed up under the inherent conflict of interest that lies at the heart of for-profit provision of care – that is the contradiction between providing the best of care and maximising profit.

Where profit becomes an objective in itself care is bound to suffer, there can be no avoiding this point whatever finesse is thickly spread by marketing or "brand building". This conflict of interest has only become sharper during the current - and in terms of the rising cost of care for an increasing number of ill, elderly or infirm - enduring funding crunch. Evidence of profit vs care can easily be found everywhere in adult care - from the cutbacks on the quality and variety of food offered to service users in residential care, to the curtailment of activities and stimulation offered by overstretched CW's, to the time limited "appointments" of domiciliary carers. However it has more subtle effects too. The for-profit model is based on building facilities or services and then running them at the extreme end of the their capabilities – for-profit care depends on "capturing" and "retaining" service users regardless of any limiting capacity. This produces the most adverse of results and include such failures as the placement of service users in wholly inappropriate care settings exposing them to fear and danger - not from CW's - but from other service users and which degrade and diminish the quality of their lives; the denial of services either through a reduction in staffing levels or "gating" basic services – that is making the most ordinary of functions associated with good care supernumerary to the underlying cost so forcing families to pay for services that nominally should be part of any adequate care package; the unnecessary suffering of individuals by overworked, inattentive or disinterested clinicians; and the exclusion of families from care decisions as a method of preventing proper oversight of failing care.

Perhaps most serious result of this "capture" and "retain" policy of care is the part that private care providers play in effectively defrauding the taxpayer – if not in strictly legal terms then at least in terms that fly dangerously close to that definition. In order to maintain or increase the revenue streams from service users I have witnessed families being "coached" through the application for Continuing Healthcare - including the manipulation and fabrication of clinical evidence to exaggerate a service users' needs so that they can meet the appropriate criteria. The advantage of this for families is that, and I quote from the NHS Choices website –

"NHS continuing healthcare is free, unlike social and community care services provided by local authorities for which a charge may be made, depending on your income and savings."
http://www.nhs.uk/chq/Pages/2392.aspx?CategoryID=68
The benefit for families is clear – they are not means tested for contributions to care. This though is not an altruistic exercise, as for-profit companies, once they have a Continuing Healthcare service user on their books, are able not only to better "retain" that service user - who may otherwise have been moved to another provider that offered cheaper care - but they are also able to milk NHS funds for all manner of additional care charges (such as overcharging for care fees, board and accommodation) that they would not have able to if the service user had been reliant on their families to pay for such care. By this method I have seen families with 7 figure resources obtain full funding for care costs at the taxpayers' expense.

Of course none of the bounty of this NHS "largesse" ever finds its way back into quality of care provision which remains, as outlined in the above, under unremitting budgetary pressure, leading to the obscenity that NHS funds, at a time when even these are stretched, are diverted straight into the pockets of for-profit service providers.

This effectively makes a nonsense of the one supposed advantages of the free market in care – the fact that competition will hold down prices. In fact for-profit care has been the single biggest driver of increasing healthcare costs for the elderly, ill and infirm over the last 20 years.

All these are just the problems being unmet or unseen on the, in economic terms, demand side – that is services failing those they are meant to serve.

On the supply side – particularly on that of CW's - the issues are even worse. CW's are treated overwhelmingly as infinitely replaceable units of production; they have no intrinsic value as individuals, whether they are good or poor carers, and if they show the slightest degree of free thinking or – God forbid – questioning of practices, they are dispensed with easier than biological waste. Bullying and intimidation - overt or covert – of CW's by service or home managers, senior carers and nurses is not just commonplace but the norm, with threats of sanction and disciplinary's against "troublesome" staff summoned up out of thin air supported by "evidence" supplied by other suborned CW's and clinical staff – instances of which will be shown in the following pages.

Just like a fish, the rot coming from the head extends downward in creating a culture of intimidation and bullying among care staff that is hard to fathom in its depth such is the brutality, viciousness and vindictiveness that goes beyond the petty to the deeply cruel and is worthy of the name of abuse. This is often coupled with obdurate racism that permeates all levels of care and is often accompanied by language often associated with 1970's attitudes toward other cultures and nationalities.

At the same time more and more pressure is being heaped on fewer and fewer carers as service providers are seeking to cut costs (to maintain profit margins) in the financial squeeze that care has been placed under by recession and austerity. CW's are operating at the very limit of what they can do safely both in terms of the hours that they are expected, or asked, to work and the care that they are asked to deliver when they are at work. This overstretch is manifest from the heroic levels of sickness that permeate care - and are the first sign of an overstretched and overstressed workforce – to the failure to perform such basic tasks as feeding and hydrating service users and on to the chronic turnover of staff that often leaves inexperienced and under-trained predominately young or overseas staff fulfilling roles they simply are not adequately prepared for or capable of.

Carers now have no time to care and as a result service users suffer – when carers have no time but to attend to immediate tasks such as personal care and service user hygiene (bathing, showering, and washing) service users are left un-stimulated, unoccupied and bored, often provoking challenging behaviour that would not present itself if they were given meaningful or enjoyable activities. This is not to mention the issue of poor or failing staff. Many of these are carried by good staff who therefore come under increasing stress as they cover for the mistakes, oversights or laziness of their colleagues leading to a drain of good and experienced care workers to another care sector or out of the profession totally. Despite such failings poor staff still manage to be retained either by reasons of favouritism, the inability of managers to see what should be largely manifest or by the loss of good staff leading to difficulty in recruitment of replacements.

In varying degrees all these issues have so far been papered over by an active and engaged voluntary sector which has for some time offered activities, stimulation and basic care that has not been able to be picked up by paid care services. Even these though are now withering on the vine as cuts in funding have forced many voluntary groups to close for want of help paying for the renting of rooms or covering of incurred expenses. The loss of this sector of care leaves the poor state of paid services exposed, but the burden falls not on service providers but on those they serve. Adults in care are increasingly left alone and unengaged either in their own homes or in residential facilities and are condemned to a life of recurring boredom, depression and despair. I know because I have witnessed it and, in some cases, been a party to it by implication.

The sum of this is what has driven me to write from the personal and professional standpoint. It was either this or to leave care altogether which I was, and am, reluctant to do. What I want is for people to wake up to the fact that as a society we are implicitly condoning the failing nature of care, we are standing by while the cost – the real cost - as in quality of life and variety of existence – fall disproportionately on those least able to do something about it. I do not want to be part of a care system that takes an active role in the diminution of life opportunities for the few that most of us take for granted.

To paraphrase Hemmingway, care is a fine job and worth fighting for. Right or wrong I take my stand here.

PART I

The Dark Wood

And I gave my heart to know wisdom, and to know madness and folly: I perceived that this also is vexation of spirit. For in much wisdom is much grief: and he that increaseth knowledge increaseth sorrow.

Ecclesiastes 1 vv17-18

CHAPTER I

Who Cares?

In my time in care I found that Care Assistant's (CW's) - while often being an eclectic bunch of people all from different and diverse backgrounds and with wildly fluctuating levels of educational achievement – were united – without exception – by falling into one or more of three broad categories that drew them into caring – what I would call **The Three D's.**

First are those who end up working in care almost by **default,** they have no particular affinity for care work but their lives just seem to have led in that direction (a terrible reason), second are those who think a job looking after people less able bodied or more mentally challenged or sicker or older and more infirm than themselves is a good thing in itself, that it would make a real difference to the lives of those they care for, and in so doing make the world a better place, (an even more terrible reason) these I would term the **deluded**; and thirdly were those who ended up in care because in their lives and backgrounds they had missed out on a loving environment for one reason or another, and they believe that by giving to others the love and care that they had never had themselves that they would too can become lovable to others (the worst reason of all) – the **damaged.**

I am not being unnecessarily cynical when I ascribe to all these motivations (or lack of them) various degrees of terribleness because in fact in the largely unregulated field of CW's - because it demands no professional training or preparation, because there is no mandatory qualification that has to be attained before entering any workplace and, because once CW's are in a job there is no proper nationally recognised method of oversight, monitoring or evaluation - then the whole process of who works in care and who doesn't is left largely to inadequate and cursory interviews and often flawed references; what this means in practice is that different people's motivations for entering care becomes ever more paramount in both how they deliver that care and how they progress as CW's.

Neither am I being patronising in describing care workers in such a reductionist and arguably critical fashion – remember I too worked in care, and, as I noted that there were precious few exceptions to these three categories, then I too was just as flawed in my motivations in entering care work as all those others I met.

This is like a version of the Liar's Paradox,
http://en.wikipedia.org/wiki/Liar_paradox

Where the contradiction in the statement – "This statement is false" leads to a logical contradiction. All the following comments I make I make as a CW myself, therefore I am prone to all the flaws and failings that the **The Three D's** make manifest.

I'mWhat is important in divining these categories is the fact that what they represent is counter-intuitive to what many laypeople – those not intimately knowledgeable about care – think the motivations are behind carers becoming carers – that they do it because they have a vocation for it – this is almost without exception wrong, carers end up working in care because they need to fill something in themselves.

So what, you may ask – the same may apply to any job, particularly in the caring professions and no-one is suggesting it be a bar to say a doctor or a nurse becoming one.

That much is true until you consider that CW's perform the grunt work of care – they clean the incontinent, they feed and hydrate those unable to feed and hydrate themselves, they wash, dress, comfort, stimulate and support those who they care for and also monitor them for signs of illness or disease; it is then that an almost total reliance on hazy notions of motivations why certain individuals enter into care in the first place and how they monitored thereafter becomes critical and the lack of it even more shockingly unsatisfactory.

As a result all manner of ills are (both fairly and unfairly) attributed to CW's and, as they have no professional body to either keep their feet to the fire through sanction, or to defend them against unfair accusations, and because there is no proper system for ensuring bad CW's are barred from further care work, then you have a work environment somewhere close to the wild west where the fastest gun gets to wear the sheriffs badge.

In the following pages I will go on to show how all manner of dangerous situations have been made almost critical because CW's failed to either do their job, be properly directed or were actually a danger and liability to those they were supposed to be caring for. Also though in the following will be plenty of examples of CW's doing fantastic work under the most trying of conditions and often against pushback from poor or incompetent clinicians. The point being there is no way to root out bad CW's or to recognise good ones other than through the often the capricious choices and decisions made by mangers or contractors of a service. In this light it is less a shocking state of affairs that some CW's behave appallingly and more a wonder than such catastrophes such as Winterbourne View and Stafford Hospital don't happen more often (in fact, as I shall go on to show, Stafford Hospital may in fact be only the tip of the iceberg).

These though are subjects for more expansion later, first we need to go back to those reasons why people go into care – key as they are for CW's aptitude and ability - in order to get a proper appreciation of how differing guiding values are funnelling into the care system people who are wholly unsuited to working in a caring environment.

Those who wind up in care by **default** are most often young and female and with less academic qualifications than the average and often without a clue what they would actually want to do with their lives; instead of a motivation toward care, they have been prodded by careers advisors in school, based on their poor academic prospects, in that direction. Female and with less ambition and aptitude than others? They must be good at caring right? They are women after all and don't all women hanker after looking after someone or something? Ergo care work is for you.

This matters for two obvious reasons, first, as far as academic ability goes, not every CA should be expected to be a sleeping rocket scientist or neurosurgeon, but when dealing with a line of work where paperwork is often next to God in the rank of importance then a certain level of literacy is needed, so too is numeracy – for example how would a CA know they have properly hydrated an individual if they cannot add up properly the fluid intake recorded over the course of a day?

[N.B. It is important to note here I am not talking about intellect, or a lack of it, which is of no consequence as to the making of a good carer – smart or not so smart, intelligence has nothing to with caring well - I am talking about real and severe learning disabilities, there is a world of difference.]

Secondly, lack of numeracy and literacy aside, what does this method of directing individuals into care say about the value we as a society put on care if the only criteria for providing labour for it is that all other career options have been exhausted? The downside to this process is obvious, the healthcare field on the non-clinical side is dominated by individuals who never envisaged themselves as carers and often times have no appreciation of the importance of their work or the concomitant skills or motivation to provide the best quality of service possible, for here we are not talking about processes but people – a worker on a manufacturing production line makes a mistake and a product is ruined or defective, a CA errs badly and people are exposed to the most awful suffering, discomfort and, in extremis, harm.

[N.B.It should be noted here that many different reports – the Cavendish Review and the Kingsmill Review to name but two - both of which I will draw and remark upon later in more depth – have commented on the average age of the workforce in care as being "mature" - that is around 35 years of age. However, as a figure also quoted is that almost half workforce is 46 or over, this means that there are a very large number of very, very young CW's in order that the average is squewed downward.

It has been also noted that the profession has trouble attracting young CW's but with the cutting of working age benefits due to austerity, most for the under 18s, more and more young, mainly female unemployed people are being "driven" into care by the processes outlined both above and below.]

Even allowing for the many excellent CW's I have worked with who also "accidentally" have wound up in care this method of creating a care labour force still produces a lottery that no one would want to be on the receiving end of; would you want the welfare of your nearest and dearest to be passed into the hands of a largely very young workforce who have ended up in care because no alternatives were available? What's more this is a terrible waste of talent – a lack of academic ability does not mean a lack in all attributes, by funnelling those with no real appetite for care into the care field it cripples the chances for those individuals to do something they genuinely find stimulating or interesting and would much rather be doing.

If it is possible though things are even worse, and problems greater, than simply the employing of people with poor numeracy and literacy skills; there are actually CW's – and I have met many - who not only are academically disinclined but have actual Learning Disabilities themselves, some quite severe.

In making this point I am not being discriminatory, everyone has something to contribute to the world of work and frequently I have found my colleagues who do have learning disabilities make up for their lack of aptitude in certain areas by excelling in others – an individual with learning disabilities may lack basic literacy but still be a very good carer – however I have worked with other individuals who not only lack the academic basics but who have real difficulty assimilating instructions and information to the point at which information that needs to be acted upon is left undone, needs are left unmet and damage inflicted unnecessarily. The situation often arises – and I am not exaggerating here – that some carers themselves require the support of care professionals, in other words the very carers need caring for.

Once more this is a horrific situation to place vulnerable people in as it not only exposes those that are supposed to be cared for in danger, it is also unfair on those carers with learning disabilities on whom pressures and responsibilities are heaped that they are in no way prepared for. This leads to another open ended question – how many failures in care – resulting in real harms or even deaths - have been put down to poorly supported carers who are themselves in desperate need of care to live and function successfully and who have been effectively scapegoated in the aftermath of a mistake or tragedy?

Whichever way you look at it this is some way below a satisfactory level of care provision.

And the method of "choosing" CW's arbitrarily does not end there, many carers I have met and worked with were directed into the profession by the Job Centre - again because of a lack of alternatives - and this exposes perhaps the biggest and most dangerous gap in the care recruitment process – that of squeezing bodies off the dole and into a work environment that should be far better regulated.

It works something like this - a provider may be opening a new service, so they advertise and also inform the job centre of vacancies, the job centre then drags in the young, under qualified and demotivated and packs them off for an interview, regardless of whether they have expressed a desire to go into care or not; the service provider, aware that revenue streams will only flow when a service is up and running, are wholly indiscriminate in who they take on, so it is that someone can make the journey from dole to care in a single push. Further, employers have an incentive to employ the very young into a service. Most employers pay only the minimum wage - currently at the time of writing £6-31 - however for those under 20 years of age it is £5.03, and for those under 18 £3.72, therefore the dice are loaded toward employing the youngest staff possible because of their cheapness. As the minimum age for working in care is 16 you have a situation where employers will be financially interested in employing the youngest people possible. Employing people with no desire or aptitude to work in care but with some maturity is bad enough, employing similarly unmotivated individuals who are still basically maturing is even worse – you end up with a workforce who are wholly uninterested and vastly too immature to understand that their actions – or more often inactions - have real consequences on other living feeling beings.

Care is tough environment to work in and frequently demands huge effort over long shifts, if we take the received wisdom that money – and how much money – is the main driver of the will to labour we have to ask – how hard would you work for £3-72 an hour? Could anyone really blame a 16 year old for not exactly stretching themselves in a job they never wanted?

It is obvious how risky this method of recruiting a care workforce is – although CW's, or prospective CW's, are screened, first by running the names of prospective employees against the Safeguarding of

Vulnerable Adults (SOVA) register that lists those who have previously harmed vulnerable people in their care, and second by applying for an enhanced Disclosure and Barring Service (DBS, although often still referred to as CRB check due to the previous registering body the Criminal Records Bureau) check that lists all past criminal convictions and unspent cautions – this screening is less that useless because many of the recruits have never worked in care before therefore no red flags will be raised through the SOVA check, and as the DSB disclosure is only advisory - it tells an employer about any criminal history it does not make recommendations as to their suitability to be employed working with vulnerable adults - it is ultimately up to the discretion of the employer if they go on to recruit an individual. Once more it can be easily imagined that an employer with posts to fill will more likely turn a blind eye to anything but the most serious criminal infractions.

[N.B. The suitability of employees as regards their criminal histories is a core standard of the CQC inspection criteria and something they will look at in any audit or inspection but, as will become apparent later on, the CQC have neither the time nor the resources to check every single care employee that works in the private sector or even in the public one so once more there is an over reliance on employers to ensure that they are not employing people with pasts that may be indicative of harmful or abusive behaviour in the future, a reliance that is undermined by the conflicts of interests with employers that have been highlighted above.]

Add all this together and, as I noted above, disasters in care appear not as aberrations but as the inevitable result of a flawed recruitment process; care has become a basement line of work. Poor pay for immature, unmotivated and ill-supported staff who frequently don't want to be there and who include a significant minority who should be recipients of care support themselves produces all manner of perverse results including not just avoidable incidents but stratospheric levels of sickness, horrific levels of bullying and a work environment that is totally dysfunctional, but these are all subjects I shall return to in due course, for now, we turn to the second reason for people going into care – the motive to do good or those who I have called the **deluded**.

In using the word "**deluded**" I am not seeking to classify such carers who meet this definition in a pejorative way, it is only, as the following will make clear, the most concise way of describing individuals who either impose impossible pressures and ideals on themselves that no one could possibly cope with, or else it creates perverse care outcomes that in fact operate to the detriment of those they care for.

It would seem odd at the outset that caring too much is a flaw in an employee in the care field as this would seem to be an essential requirement, and so far as it goes this is true – a good carer will be someone who is diligent and motivated – something that a caring spirit can enable in the face of the numerous obstacles that are placed between a carer and those they care for. Yet caring too much, and for reasons of believing that through such care that everyone that they care for will be happy and content is wholly unrealistic. Caring because you believe that you are doing an essentially good thing is like what someone once observed of the legendary independent music label Factory Records – it cannot be long sustained once in contact with the real world. Care is messy business – there are no neat bows, no tight corners, and, most importantly in this context, no end to it, a carer can spend the whole of their 12 hours shift (shifts in care are typically 12 hours long – something else that is also fundamentally flawed but more of this later) doing nothing but caring – pausing neither to eat, drink or rest a carer can lay everything on the line – and yet at the end of a shift there will still be things left undone, still be more care that the carer could have given and, perhaps most important to someone who cares too much, some of the people you support will still be unhappy with the level of attention they have received. Although painting the Forth Road Bridge is an awkward analogy in respect of care, it is still valid - only this apocryphal Sisyphusian task as it applies to care should not just be restricted to painting the bridge in question but wallpapering it too. What typically happens to those who care "to make a difference" is one of three things, all with the same result – they burn out, they get taken advantage of or they end up making the serious mistake of believing that what they perceive as care is what the person being cared for actually wants or needs – the endgame of all of them is that such carers are either forced, or "fall", out of care.

Burn-out is the most obvious way in which such carers are brought to their knees – the human brain and body can only give so much of itself, only do so much caring, in the end emotionally, psychologically, physically and mentally there is point where the brain or body calls a halt – stress, depression, serial colds, flu's, Diarrhoea and Vomiting

(D&V) bouts and chronic tiredness – all of these are symptomatic of burn-out, what's more, those prone to caring too much are also the most unlikely to seek help for their condition as they often feel that to admit that they cannot keep up this level of caring on an emotional, physical, mental or psychological level is, to them, an admission that they have failed somehow. As result their effectiveness as a carer is blunted and in the end arrested, they die a death of a thousand cuts – too ill to work, too caring to take appropriate time off, eventually they run out of road and are destroyed totally as CW's – mental breakdowns are made of such stuff.

All this is almost inevitable in those who care too much, and this in itself would be enough without the additional factor than many carers have a keen nose for those who care with all their fibre – and are only too ready to take advantage of the fact.

A friend once said to me that care is divided into carers who don't give a shit and those who clean the shit up after them. They were largely right, only their observation didn't put into context the numbers or ratios of those that care and those that don't; worryingly for anyone with a relative in care is that those who don't give a shit far outnumber those who do. The upside for relatives is that those that do care keep the whole ship afloat, whatever the majority of carers lack, the minority makes up for them, and this is a major point that needs emphasising here – **most good care is maintained by a very few good carers who do their job - and more - and which consequently leaves the service in which they work relying on them for the continuance of good care.**

This obviously leaves good care on a knife edge, unbalanced and disproportionately weighted, all it needs is for one or two of the good carers to leave and the service will go downhill rapidly. This is why many failings in care homes are perplexing. A CQC report may give a home a glowing report, yet, within 6 months, or even less, the same home can be in the news for neglect, abuse or some other failing – this is directly down to critical good carers leaving; and good carers do leave and well before any bad ones. There are several reasons for this – from the obvious – poor carers would have trouble getting another job – to the bizarre – that good carers' work is seldom recognised and appreciated by the management of the service in which they work – but the main one is that they simply get sick of picking up the pieces after poor carers. For good carers find themselves in a vicious circle, they clean up the mess left by bad carers, which doesn't go unnoticed by the bad carers, who leave more and more mess to be cleaned up the by the good carers because they know they will be too conscientious to leave it and so on and on and on. Eventually good carers get fed up and leave, only they find little relief, the reliance on a very few good CW's is sector wide, no matter where good CW's go they will find themselves in the same situation; taken advantage of over and over they - as with burn-out cases - get used up and eventually leave the profession altogether.

Caring too much also has other more baleful effects that are less obvious but just as critical - for both the carer and those they care for – such as the imposition of care *that the carer believes is right* rather than what is appropriate for the person being cared for. Richard Sennet the Centennial Professor of Sociology at the London School of Economics and University Professor of the Humanities at New York University wrote in his 2003 book "Respect" about the need for some form of reserve in caring, in short a capacity to care without caring too much, as he puts it, quoting his social worker mother, caring should have limits and boarders –

"To make compassion work perhaps it [is] necessary to defuse sentiment, to deal coolly with others. Crossing the boundary of inequality might require reserve on the part of the stronger person [as] reserve would…acknowledge a signal of respect."

In care you need to care to the extent that the needs of all are met equally first and foremost, you are there to enable those you care for to live as best they can on their own – often very limited – terms; to care too much is to deprive the people you support of the feeling that they are equals with you, that although they need support they do not need nannying, to do so is to make the cared for feel emasculated, treated like a child, insulted. Yes caring is important for empathy but more important is that you meet *their needs* not *what you think their needs are* – to do so you cannot place yourself between those you care for and their autonomy and independence no matter how halting or proscribed that autonomy or independence is. To do more than care professionally – at a distance marked by reserve – is to create dependence – this is not care, it is infantalisation – the reduction of mature human beings to the status of a child.

Carers who care too much are too often prone to this form of dependence creation, it is unhealthy for them as it diminishes them as carers – and unhealthy for service users as it reduces their potentialities and stunts their existence. In care you need to care enough to do your job, but not so much that you start to make up your own rules as to exactly what that job is.

The last of **The Three D's** we turn to is those in care that have entered the profession in order to heal something shattered or broken in their own backgrounds – the **damaged**. This draws a surprising amount of people into care and can have both positive and negative effects on the care of the service users and the service that provides for them.

Care work for many people is a form of escapism, in dealing with the problems of people who are emotionally distanced from you it is possible to escape from the demands and despair of those who aren't. As such, care work can demand little from you and pay out a lot – you can seamlessly move from a situation in your own life where things appear to be uncontrollable to one where you can *feel* that you are in control [N.B. The illusion of having control in a care environment is false; it is a manifestation of the Illusion Of Control, an important factor to which we will return later]. When I entered the care field I was see this in action - how people who enjoyed so little care in their own lives, whose own needs went unmet - were able to immerse themselves in the giving care of others; in meeting the every need of strangers they sought to find a little salvation in process.

Such individuals – those with broken or dysfunctional backgrounds – made neither exceptionally good carers or incredibly bad ones, in time I was to observe that those who went into care work to escape from their own lives were neither predisposed to be well fitted to their profession or should by rights be excluded from it, the only thing that really marked them out was their sheer numbers. There were those who had been abused as kids, made homeless as adolescents, beaten by possessive boyfriends, betrayed by husbands to drink, other women or other men, had brothers or sisters on the needle or the stone or the lam, and appeared to have offspring of an age that only led me to conclude that they must have borne them as almost kids themselves.

Such dysfunction, while not exactly defining them as carers, did have consequences at work, on the plus side such chaos gave such individuals every reason to turn up for work – to escape the domestic horror – but it also meant they had regular and legitimate reasons for being off for erratic periods, this made most services I worked in run on a knife edge of staffing. On almost every shift there was one or more carers missing, meaning the pressure was added to those who turned up – and given all the other pressures of work, and the fact that a majority of those turning up for the shift had their own dysfunctions to deal with – all this just ratcheted up the tension in the care environment; not an ideal way to support ill or vulnerable people. The one thing that always stood out though was, when such individuals turned in, they could almost completely disengage from whatever was going on in their personal life, they could leave whatever was failing in their own lives at the door and switch on to work. This was something I partly envied, I could not do that, but perhaps that was because, in comparison to those I worked with, my domestic arrangements and background was idyllic to the point of halcyon.

Such were the people and backgrounds of those I came to work intimately with – the accidental, the activist and the agonised carers. I neither pitied or patronised them as the first thing I found in care was that watching your own back was the first rule of survival – a lesson I learnt, then, critically forgot, to my major cost.

One final word here has to be added of which no explanation of care and those who have ended up working in it, is complete, forming as it does the majority of care staff - and that is the number of immigrant or emigrant employees that are found there.

Go into any care environment, but most especially one in the private sector, and you will find yourself in something that resembles the United Nations. Almost every extremity of the globe is captured in care work from such diverse backgrounds as Pakistan to Russia to the US and most especially Africa. In my time in care I have worked with Filipinos (the current racial group de jour at the time of writing, although there are in rising numbers of Chinese and Vietnamese – probably the boom group of the next wave if immigration rules on non EU members are slackened, Romanians being the default option if restrictions remain in place) Indians, Afghans, Thais, Yanks, Poles, Slovenians, Iraqis, Nigerians, South Africans, Pakistanis, Fijians Nepalese and a few that may have slipped my mind or been unconscious of their actual nationality.

There are numerous reasons for this and many more myths; whatever – either reasons or myths – all have uncomfortable truths at their basis that will anger or inflame nearly everyone, however they *are truths,* as I have personally been witness to; and however uncomfortable, avoiding telling them is even worse than not discussing subjects because they are supposed to be out of bounds.

A broader and more in-depth discussion about immigrant labour in the care workforce will be covered later, here though it would be remiss not to mention three points as they apply to the chapter heading question "Who Cares" –

First - that immigrant labour is used in care to artificially hold down wages. Without large scale use of immigrant labour the law of supply and demand - because not everyone is either inclined or cut out for care work and few find the idea of being beaten up by septuagenarians (it does happen) or mopping up excrement (it always happens) a reflex career choice – would have inflated wages beyond the minimum at which they are currently fixed.

Second - care providers, even medium sized ones, actively recruit overseas workers by opening offices in far flung nations (one care firm I worked for had an office in the Philippines) that interview workers, arrange visas, sort out UK accommodation for them and guarantee them work for the length of their visa stay.

Third - overseas workers are vastly more reliable than native UK workers, work more and harder, take fewer sick days, and, in most cases, are far superior in the care they provide.

Because of these three basic facts race and racial issues are one of the most contentious issues in care. Nowhere else, not even in the most roughneck of industries I have worked in, have I encountered racism on a scale like I experienced in care, and I mean open racism of the sort that would have thought to have been long banished among civilized people. Yet this is small wonder, take a dash of white "low skilled" (the whole term of "skills" is another barrel of fish to be shot at later) poorly educated low paid and largely working class labour, give them a job that eats away at their self esteem and demotivates them at every turn, throw in a combustible mixture of vulnerable but often violent mentally ill clients, a nip of private owners who's whole focus in providing care is to turn a profit at any price and then boil the whole thing up with a racially mixed cocktail - often in places with low racial intermixing - and you are virtually guaranteed to have "problems" and the bitterness of the brew is corrosive.

Who cares is one of the most basic questions that is never discussed in all the political arguments over how care for the elderly or mentally ill is delivered, it is as if who actually delivers that care are unimportant. This is so obviously totally wrongheaded it hardly needs pointing out here; you can throw millions at care right now and little of it would stick simply because the composition and nature of the workforce who care is given so little attention.

Here is not the place for recommendations or conclusions but it is impossible to pass by without noting that the whole structure and nature of the care workforce needs to be overhauled or at least thoroughly examined. Some attempts have been made such as the Cavendish report which I will go on to discuss a little later, but this had been largely a tinkering at the edges. To my uncouth and untutored eye this is like the owners of a football club lavishing money on a state of the art stadium and facilities but leaving the actual team untouched, then when performances are routinely poor spending yet more money on the stadium and the facilities and scratching heads wondering why the team isn't winning games. Unless the team is appropriately skilled, unless it is motivated, unless it is properly rewarded and most of all unless it is properly formalised through a professionally body that will be both representative and standard setting nothing will change – you will have a world class stadium and facilities and a Conference North team.

For all the seemingly dysfunctional nature of the workforce that I encountered and for all the implicit criticisms I have levelled it is also ultimately true that there are very many excellent carers out there, but they are woefully outnumbered by the poor ones and woefully unsupported by the structure that surrounds them. Until they are recognised and more is done to reconstitute the care workforce the crisis in care I outlined in the above will not just continue but get worse.

CHAPTER II

An End, A Start

This is not a book about me, it's about care, and it's not a confession either; yet it is also all about me and a confession – for how could I mark down my observations of care work - and where they lead - without filtering them through my own personality and make up, and how can I talk honestly about that personality and make up and make assessments of truth about care work unless I can confess to my own limitations and fallibility as a carer.

To those who think I have been unnecessarily tart in my observations of Who Cares, let me remind them that I too am a carer and therefore according to the categories I outlined in the previous chapter I would have to say that I don't suppose I fall easily into any of them, and this is because a little bit of all of me falls into each one – I didn't ever plan to work in care and ended up there by default by way of death and grief, so I was an accidental carer; I did have illusions and delusions of "doing some good" in care – something I never quite fully shook off in all my time working in the field, despite it taking on a cynical and self-preservationary turn; and my background, though compared to most, was privileged and loving there was also a part of my history that was every bit as chaotic and dysfunctional as that of those I worked beside.

The plain fact was that ended up in care - by way of a number of other jobs and stalled careers - through the death of my father; the Panglossian view would probably look on this as something good coming from something bad, that there are no real adverse events, only a failure to spot opportunities in them for self-discovery and renaissance; I humbly submit that this is bull, although I am open to the fact that I probably would not have ended up in care without the death of my father and the manner of his passing; yet even this apparent opening was something of a curse, no job has challenged me more than care, no labour has stretched my mental reserves as far as care, and I have had more sleepless nights (or days) through it than any other paying work ever has.

My life and work history was a disaster on paper and worse in fact. It started off promisingly enough, I was a reasonably bright kid and left school with a couple of A levels, then I worked for a year in forestry before going to college and studying it for two years. It was supposed to prepare me for a management job but on finishing my qualification I didn't fancy life in an office; no matter how many times you could ride out and into the woods you were supposed to be managing, there was nothing to substitute for the smell and freedom that can be found working in the open air.

The other thing that put me off was the thought in ordering people about, telling others what to do, most especially telling working men (and it was nearly all men) who worked hard that they had to work harder; to me - rightly or wrongly - managers were people with soft hands who lived on the fact that they once did a bit of work out in the sticks, down in the trenches, and which time had come to greatly overestimate as to its length, difficulty and struggle and yet who, on that limited basis, always saw the fact that more could be done with less.

I had had some experience of this, when we were all sent out of college on assignment, and I absolutely loathed it – I felt like a charlatan, and I saw too the resentment in the eyes of those I was critiquing about the job they were doing - the narrowing of them, the distain in them, the absolute lack of respect in them. I suppose I was a coward, I wanted to be the guy out in the sticks, I wanted to be the guy in the trenches, and if there were any resenting to be done I wanted to be the one doing it. The advantage of being at the bottom of any pile of bodies is that you can always look upward and see only arseholes. I guess that a psychologist would see some inferiority/ superiority complex going on, the fact that without responsibility you could live without consequences – a dumb idea, but to those who have it a comforting one. Besides, I comforted myself, management posts were difficult to obtain in a climate at the time marred by recession, when the forestry industry was in flux and when the Forestry Commission – once a rich source of management jobs - was selling off land and outsourcing work, so I went into self-employment as a tree harvesting contractor – or "cutter" - felling trees with a chainsaw on piece rates and living on site in various locations in a caravan. This was a semi vagrant life as you don't get access to much electricity in the middle of a wood nor running water, never mind hot running water. Still I probably enjoyed this job more than any other I was to do, but it couldn't last. It was in fact living a life denuded of common luxuries, without a TV or other distractions that marked the end of this passage of my life as I started reading books to pass long winter nights and developed a passion for literature. Eventually I enrolled at a night class and got an A level in English and applied for university and, to my eternal shock, I got a place and so packed up my semi-bankrupt business and went to study. It was here I learnt the first real lesson of my life – if you enjoy something never do it full time. I enjoyed some of the work but for the most part I struggled to plug into university life – never the most socially adept individual and being a good 10 years older than my cohort group there was little chance of fitting in. In fact I had never really quite fitted in ever, I was one of life's outsiders, never having the feeling of being "in" and always seeming to glance obliquely at existence. To make up for this disconnectedness I drank – a lot. A second lesson soon arose that took me some time to learn - you cannot short cut *being* and certainly not by substance abuse – the more I swallowed alcohol (and other things) the more it swallowed me, eventually I sank and ended up in a locked ward at the local hospital psychiatric unit with an unfinished degree. I can't say if I was screweded up before I started drinking or because of it and to me it doesn't really matter, by the time I quit drinking I was on some pretty heavy medication.

Even without alcohol, or maybe because of its absence, I developed deep black moods that landed me a couple more times in a locked ward, each time I emerged with more medication and less of myself, although it didn't matter because I was lucky in my misfortune, I ended up in a fellowship that gained me the best friends it was possible to ever have and probably saved me from the rope – a fate that befell many of my previous friends and at least one relative. Eventually my medication and I reached a truce where I could function and work, so I went off to work on a market stall, then a supermarket and then on to the kitchen assistant job at the local authority Civic Centre. Here I stayed for perhaps the longest I have ever stuck at work since my wood cutting days. Still character flaws and self-destructive tendencies found me out once more,

I screwed up by trying to help someone else out – Lesson 3, trouble always comes when you consciously try to do some good – and, due to me being an idiot with no clutch pedal between my brain and my tongue, got stuck on the wrong end of a disciplinary.

Once more I was lucky as I had joined a union while I was there and they negotiated a severance package – I wouldn't let my mouth go about all the little fiddles and perquisites that councillors purloined for themselves and they would cough up some bucks. This experience left me with no job and no reference and also earned me yet another trip to hospital – hopefully my last. There followed a long recovery in which I was on Invalidity Benefit and relying on my partner who I had met a couple of years previously and who I had now moved in with. It was she who supported me financially and psychologically and without whom Christ knows where I would be now. In this time I started writing and even managed to sell a screenplay that paid for a holiday – the only one we have ever had together - and it was during this holiday that I heard, by a crackly mobile phone while stood on some Cornish cliffs, that my father was terminally ill with cancer.

No one can watch the slow unpicking of the seams of an existence and not be moved in some way. You may be awed at how nature ignores nothing, how it alters everything, changes tissues, organs, bone, fluids, and makes them agents of decay. You may be horrified at the petty indignities that slow death visits upon the dying, the loss of continence, the way the face, so long loved and recalled, becomes distorted by the still living, living end. You may be surprised or shocked that when it comes death it is not quite the way you may expect it from films; few are the times, lucky are the people, who have the Hollywood last frame, the ceasing upon the moment, a slight catch of breath, a final heave of the breast, a fixing of the eyes on some far shore or some distant horizon, a still point beyond……..and then silence. Or perhaps you may feel nothing, or if anything slightly cheated at the way death steals up without a fanfare, with only the shabby tribunes of petit mal strokes and oedematous lungs, detached you only watch the flood envelope the body.

I watched my father die slowly. Cancer had bitten off most of him, leaving the best of him to last. First in the prostate it whittled away like some tree mite, then it flowered in his lungs choking off his voice to thinning whisper, and finally it struck its roots down in his brain.

Being without a job proved to be an advantage now as I could help my mother help my father spend his last days at home rather than expiring slowly in an anonymous hospital ward.

Every week I travelled down the short journey by train to my parents' house, every week I saw less of my father. He was there of course but more and more of him was becoming less and less of him.

During this time it had become necessary to dress him, he had developed a large tumour in his hip that restricted his movement and gave him constant pain the teeth of which were only kept unbared by regular painkillers (though how much pain they killed I was doubtful). He looked down at me one morning, now the eyes were quite dull, and uttered –

"Fancy having your son having to dress you."

"You did it for me so many times, I'm only paying back what's due."

And this was true, it wasn't hard to look after my father in such a way, it was as if a circle had been joined. I got a real sense of – if not pleasure, then fulfilment in helping him do the small but necessary things that he could no longer do for himself.

The decline from there was surprisingly swift, within less than a month he was bedbound, his legs finally failing him completely. Then it was necessary to do all his personal care each day my Mother and I would clean him, give him a bed bath, change his clothes, wash the sheets. It wasn't easy to give my Mother a break when I wasn't there, and sometimes when I was, two domiciliary care staff would come in. I watched them as, with infinite care, they looked after this man who to them was a stranger, they would tenderly wipe him down, clean his mouth – by this time he was being fed by a syringe driver – and tuck him back up again – it made my efforts look, well, amateurish. I wondered about what they did, I wondered –could I? It seemed callous at the time to be thinking of a future career precipitated by my dying father, like Richard III wooing across the coffin, but the thought stuck.

It held even faster when the Macmillan nurse, having watched me clean my father's mouth, comb his hair sponge his face, said I would make a good carer.

"You ever thought about care work?"

I gave some bland answer

"If you can do all this for your own father you should be able to do it for anyone."

Shortly after this my father died, in his own house and with all his family were around him; an almost Hollywood ending apart from the absolute nature and reality of death – it came after a series of deep and shuddering breaths – what they call Cheyne-Stokes – each one seeming like the last, a pull on the air by what capacity the lungs had left then a terrible pause, then another desperate air clawing breath. When the last breath issued it was almost a surprise, all of us, all the family, expected yet another heave, but there was to be none. Death came suddenly after all.

After the funeral, after all had been said and done, I found myself not just with the grief associated with my father's passing but with the additional loss of purpose that helping to care for him had instilled in me.

Weeks after my lack of focus forced me to raise the issue of care work with my closest friend who himself had moved into care later in life - taking a total career change after being in several high powered executive jobs. He too echoed the sentiment voiced by the Macmillan nurse that if I could care in such an intimate way for my father then I could convert that experience into something approaching paid work.

"But I've got no qualifications, no experience." I opined

"Neither did I when I started out. Do what I did."

"Which was?"

"Look in the papers, look in the classifieds, apply."

So I did. Day after day I bought the local paper and day after day I applied for every post in care going. For weeks and months nothing happened, I applied and nothing came back, so I applied some more, and a few more "Thanks but no thanks" letters came back, so I just kept applying ever faster. It was just my luck, I thought, to be applying for jobs in the teeth of the worst depression in modern history. Months passed with only rejections landing on the mat so I started applying for anything, Tesco's ASDA, cleaning jobs, ticket inspector jobs, driving jobs, ten or twenty applications a week, then, one day, an envelope arrived inviting me for an interview for a care job that I didn't even remember applying for; with a charity that supported those with learning disabilities. This was first chance I had had of an interview, and after all the time that had passed in fruitless perseverance it as the scent of a one shot prospect, I felt that if I didn't nail this job then the chances of me getting any kind of job, never mind one in care, would be as remote as the distant future.

If there was something that gave me hope it was very fact of having no clutch pedal between my brain and mouth – in short I could talk a good fight.

It might sound rather arrogant for a 4 time career loser with no relevant qualifications in the care field but I knew that if I could only have a shot at an interview then I would be in with a more than evens chance of talking my way into it. I had more than a couple of reasons for this, the first was that my mouth had got me out of as much trouble as it had got me into, not through any innate garrulousness – I'm actually petty taciturn to the point of almost silence with anyone but intimates – but because I could usually craft a passable narrative from the thinnest of details; and second, by luck - as I was to learn - but only had an intimation of at that time - in the field of caring it is all about narratives.

CHAPTER III

Narratives

In care the role of narratives are a vital tool – possibly *the* most vital tool - as even, as will be demonstrated below, medical histories are a form of narrative - to providing good, focused individualised and responsive care. Without the aid of various forms of narrative it is impossible for a carer – who is typically and predominantly a stranger in other respects to those to whom they deliver care – to be able to perform the role with an effectiveness.

This is down to two main factors –one cultural and one practical.

The first is that in modern care work Service Users are no longer – thankfully – solely defined by their illness or affliction, they are individuals first and the nature of why they need care is secondary.

The second is that CW's spend far more time are far involved in intimate interventions with Service Users (such as personal care and grooming, spending time in one-to-one pastoral contact, providing appropriate stimulation and activity, comforting and calming the agitated, sharing their joys and mitigating their sadnesses – to name but a few – the range of skills this called for in care workers – who remember are notionally classed as unskilled labour - will be a subject I will return to later on) than do clinicians, therefore getting to know someone's "back-story" - as Hollywood calls history and accumulated memory briefly summated – their likes and dislikes, their fears or comforts - is critically important in ensuring that the care they deliver is directed in the right way and in the right places.

While the above factors may appear self evident they need some further explanation in order that the lay reader may be sufficiently impressed with the impact and importance that narratives play in care.

Firstly cultural factors –

For those with chronic and enduring conditions – that is those who's illnesses are not acute and therefore not in need of full medical hospital care – the very type of environment in which CW's typically operate (for the moment I will limit the carers' definition to those who work in a non-acute setting, although it should be noted that CW's - or Health Care Assistant's – HCW's as they are called in hospitals - are increasingly deployed and do much of the everyday care work in acute settings) are not defined, or at least classified purely by diagnostic definitions, in other words *they are not their illness*, which is often the case when in a hospital setting. The rise of the theory of Empathic Witnessing, (again something we will return to later but, for the sake of clarity and time at this juncture, I will only glance at here) its value in the care setting, and the need to see someone's condition not as defining them but as perhaps what could be termed a causational factor in their needing care - all be it one that provides the impetus and focus for that care – mean ever more weight is given to the full narrative of person's life.

What good care seeks to do is not to see a service user in terms of what illness has befallen them, or what it prevents them from doing something, but rather sees any limitations in an individual as a lacuna in which care should be "injected" to fill while creating positive and important boundaries to that care so that the service user can be as independent, free and respected as an individual as possible. All this is only possible if the service users' needs and requirements are imparted as part of their full life's narrative. In this process the actual diagnosis forms only part of what a carer needs to know, granted, an important part – a carer should know the pathology of any condition so they are alive to the limitations it imposes on an individual and also be able to observe any changes in that condition that may have medical significance – but that diagnosis should fit into a full narrative of an individual's history. For example it is not enough simply to be told that Person A is autistic and placed at this or that point on the autistic spectrum, they should know how autism may affect

Person A's individuality and autonomy, but just as important is to know their full family history, their education, their past care needs, any employment, and, most of all, their level of *independence*, their *capacities* and any strategies in place to maintain and, if possible to *extend that independence.*

At no point of a narrative should there be a sort of index of their *dependence*; dependence in this process is not related to the individual's incapacities but rather defined in the carer's role – which should also be set out as separate from this narrative. In the case of Person A sketched above a narrative should definitely not include - in fact should seek to exclude - anything like – Person A cannot make their bed, cannot make meals, cannot read - these are all things that come under the carers definition in their role, for example Person A needs support making their bed, needs support making their meals, needs support with literacy issues – what an individual cannot do autonomously is where a carer needs to step in, but, just as importantly, they also need to know when they need to step out.

The carers role can be boiled down – and this is just what I know from experience, it has no academic epistemology – 4 vital things –

1) What are the medical conditions I should be aware of in the service user?
2) Where has this service user's life taken them to up to this point?
3) What is my role?
4) How best can I get out the way in order that this person can function more independently?

In this final point – the getting out the way part - care is perhaps unusual, as one of the most basic questions a carer needs to asks is - how little can I do to let the individual I am supporting have as much freedom as possible.

This doesn't mean that the carer absents themselves or remains idle (although the excuse of "promoting independence" is one that is most often trotted out in care to excuse laziness or care lapses, as shall be seen in due course) or does nothing, in fact an important part of a carers role is to *observe* while not *interfering*. It is this last part that is probably the most difficult – a carers' natural reaction is to step in and help at the first sign of difficulty or trouble – in fact what is best is to see that any independent action by the person you are supporting is carried out safely and that at the first sign of distress or danger they can be on hand to support them further.

It should be noted - in passing at this point - that there obviously arises here a conflict of interests well known in care – the conflict between a service user being able to assert their independence freely and the need to see that whatever they do is done safely; for to eradicate all risks you also rob an individual of their independence, therefore a balance has to be struck in care between granting the maximum level of independence possible *as safely as possible* and outright interfering– this takes us into the realm of the dreaded Health and Safety (in my experience a much maligned area and unfairly so) practices which, however much people feel have become overly prescriptive in fact leaves a lot of grey areas in care, something that poses a constant risk not just to service users but carers too. But more on this later.

None of the above would be possible without a narrative history that gives you a feel for the person and how they have experienced life up to that point – in other words a sense of how their *character* has been formed. Such a narrative, as well as showing the major changes that may have occurred – in schooling, in work or in different care settings – including the care given by parents or siblings – as it has created the *character of the Service User* - should also include salient points that the carer needs to know in order that they may provide personalised and responsive care *as applicable to that individual.*

This is where the second factor that gives narratives their critical importance mentioned above – the practical – comes into play.

By way of example of the practical application of narratives let us return to our imaginary Person A – it's not enough to know that Person A is afraid of dogs – that could mean any number of things from phobias to manifestations of their condition and perception - however if you are told Person A is afraid of dogs *because they were bitten by one as a child*, this gives the carer an insight into how to manage a situation where the Person A may come into proximity with dogs – say just walking on the street; if a carer knows this fear is real and present and *part of their life's narrative* then they can take active steps to defuse the situation by crossing the road if possible, or, better, by placing yourself between Person A and the dog and offering reassurance (I say *not crossing the road* is *better* because it does not entail avoidance, Person A does not feel they have to be the one to leave a difficult situation but that they can be protected while being empowered to hold their course. Obviously if this causes too great an anxiety then crossing the road is preferable to them having put into situation of unnecessary elevated anxiety). This example is in fact a classic case of where care fits in seamlessly into an individual's life – and the importance of narratives – Person A is empowered to interact with society on pretty much their own terms while support is on hand if and when they need it.

Narratives then give a holistic picture of a Service User – or as much as possible in a written or oral précis - not being confined to a medical diagnosis they concentrate on what a person can do rather than their limitations – their potentialities and possibilities – and reveal hidden aspects of character growth that may not be apparent if a strict clinical view was taken of their "presentation"

[N.B. Here I use the term "presentation" in a clinical context as the overall impression and summation of a person's appearance as regards their wellbeing. I will occasionally use this word in similar contexts but when I do so I will parenthesise].

So while I was unaware of the importance of narratives in care I did know the importance of *my narrative* - that is if I was to overcome my lack of experience and knowledge in care I would have to sell myself to this potential employer, and to do that I would have to offer a compelling narrative as to why they should hire me and not someone else with a less chequered mental health and work history. In fact going into the interview I could not have been less well prepared, yet also, paradoxically, this enabled me to be more of myself. My friend had told me the key points I should be aware of – the regulating body of the care sector is the Care Quality Commission (CQC), a carer should help to maintain the dignity, respect and autonomy of the person they are supporting, and that as a care worker you should aim to provide the most diverse of choices so long as those choices are informed and, as far as foreseeable, safe........... and that was it; such was the threadbare rope that I swung into my interview on.

I was also fortunate in that the charity that had offered me the interview were less conventional in how they recruited workers (though in retrospect and through later experience – much more rigorous), due to this uncongenial approach and the faith the charity had in the rigour which it applied was probably why I had been offered an interview with this care provider where I had failed in all my other applications.

The format for the day was that there would team building exercises in the morning followed by a break for lunch then in the afternoon there was a short written paper followed by an interview with a couple of managers after which you had to give a ten minute personal presentation about yourself and one area in which you had learnt something about the person you presently are.

Despite my natural reserve I through myself into the morning's team building efforts with as much good humour and engagement as I could muster. In fact I quite enjoyed it, the team I was grouped with coped hopelessly at it and I wasn't much use however we had a lot of laughs and that seemed to go down well. The written paper after lunch wasn't too daunting either as I could take stab at most questions and was not particularly stumped by anything.

What did go most my way though was the actual interview. For the subject I dredged up was a trip abroad I went on way back when I was 16, the only real contact and experience I had had up to then with people with learning or physical disabilities. Brought up a Roman Catholic I had always wanted to go to the French shrine of Lourdes with a charity that organized trips for the "handicapped" - as people with any form of physical or mental challenges were labelled at that time. I raised the money for it – a not inconsiderable sum – by doing various sponsored events and other fundraising efforts which took me up to the total needed, and so, about a week after I had finished my last GCSE, I was off to the aforementioned fated, sainted shrine at the foot of the French Pyrenees in a large adapted coach with 20 or 30 total strangers, three quarters of whom had various disabilities of various stages of severity.

Looking back I can remember the shock at the visceral nature of disability and how it affected lives like a spider's web that spread out from those with physical or mental challenges to encompass and enmesh family, carers and medical professionals. Up to then I had had the same sort of antiseptic view of disability that most people have – which isn't a criticism but just an observation -

In our fully able worldview we like to look (down) on disability with something between heavy handed patronisation, paternalism and an ideation that those with a disability are somehow worthy of our *pity* and *compassion* yet very far from our *understanding*; we like to think of those with a disability as "cute" or "brave" and rarely contemplate how any disability may affect personality, emotional well-being and intimate relationships; but most of all we rarely observe and take those with disability to be much the same as ourselves – prone to bad moods, fits of temper, flares of anger and frustration, as well as love, interest, intellect, friendship and curiosity; whether we like it or not we almost never like to think of those with a disability as people *just like us.*

On that first trip abroad the thing that struck me most was the intimate nature and ultimate difficulty in looking after someone's personal care – nappy pads, colostomy bags, catheters – these are all things most of us would rather not think about or contemplate, yet they are what those with severe disabilities have to deal with every day. Somehow on that first trip I was shocked by all that, as if there was something somehow wrong with it. Yet it was the first major challenge to my perceptions, the overturning of the common notion of what care in fact involves – it is not about pushing a wheelchair, it is not about taking Service Users out on treats to cut price seats in cinemas - although these things are important - in a holistic sense care is about looking after every facet of life, of knowing and appreciating fully the challenges that an individual may face and acknowledging that they are both fully human and fully alive; there is no time for idealisation in care, you have to see it in the raw.

Even with those without such extreme needs there is still the other challenges of seeing them as people just like the rest of us – that is why I hate any terminology that includes the word *special*; special denotes difference and difference creates barriers, those with disabilities may have altered perceptions of the world around them and how it is ordered but that does not degrade their human faculties that were outlined above – the fact that they can have moods, that they can push barriers to behaviour, that they can often be deeply unpleasant people, as well as being kind, loving, caring, funny and fiercely intelligent - *but what they are not* is *special* - to see them as exempt or outside the totality of human character traits is to do them perhaps the grossest disservice and the utmost treachery.

Needless to say in my talk that I didn't go into all of this detail but I did take efforts to impart what I had learnt about the challenges of care all those years ago, I did say that I enjoyed these challenges - because I had - and I went on to say how I had eventually been attracted to the care sector by my experience of caring for my father.

I also had to confront in this talk – as I knew at some time I had to - the issues that may on the face of it have worked against me in them giving me a job – the mental issues I had faced and how I had been supported to overcome then. Here was my USP – my Unique Selling Point – I had been on the other side of the care equation, I knew what it was like to feel totally disempowered and to need the help of others to just get by – in other words I took the most negative against me and tried to turn it into a positive – all very hackneyed, all very twee, all very successful as it turned out.

I came out of the interview knowing I had done as much as I could, and I thought I had made a decent job of it. Anyone who has been to any job interview will know what it is like to feel the interview room door close behind you and either feeling you have done ok or be filled with the sensation of it all having headed south – I had encountered both before and in this case it was most definitely in the former category.

I thought, leaving the room, that if I didn't get this job then care work was probably not going to happen for me, indeed a bleaker thought that occurred was that if I didn't get offered *this job* I severely doubted how I could get *any job*.

Two days later they phoned me to offer me a post. Stunned for a moment I didn't know what to make of it, I was caught between two poles – I was both excited and relieved yet I also was filed with a horrible foreboding – I wondered what I had let myself in for – after so long looking and waiting for an opportunity in care I suddenly panicked over if I was up to it, if I could actually not just talk the talk but walk the walk, if I had made the latest and most terrible mistake of my misbegotten work life or was entering a bright new future.

To this day I can't honestly say which it turned out to be – so much of it seems now like a horrible misstep, yet much of it also fills me with happiness and wonder.

Before I could get a start date I still had some formalities to fulfil - I had to more to fill out what was then called a CRB (Criminal Record Bureau) check form [now termed a DBS check – a Disclosure and Barring Service check that is more comprehensive and inclusive than the old CRB it has replaced] and give some other details; my prospective employers said that they would put in motion the POVA (Protection Of Vulnerable Adults) checks and that they would get back to me with a start date. Being naturally wary I wasn't going to be truely happy until I got my first day on the job under my belt, but I wasn't quite prepared when they phoned me on the 17th December – a Friday – and asked me if I could start on the Monday.

It was typical of me – three or four years out of work and just as everyone else was packing up to leave work for the Christmas holidays I was going in to my first day.

CHAPTER IV

Inductions

The flaws in the screening of potential CW's have already been touched upon but the ease with which I entered the profession is a clear example of this in action. By rights much of my history should have disbarred me from working in care or at least have given pause; one of the first shocks I got in starting care work was that the people I was supporting or caring for were often on fewer and less powerful medications than me, the fact that they had similar conditions as myself, or at least were clinically treated in much the same way, made me ask myself serious questions about what I was doing in the field or whether I was fit to be there at all.

These were obviously not questions I should have been left to square with myself but someone else should have been asking them and making a judgement from there. Yet this was not the all the fault of the "system" as it exists. Although I had been partially honest about my past with my new employer I had not been as frank as I have been here in setting down my background in black and white and at the remove of anonymity; however whilst I was asked as part of the interview process about any general conditions I had or was presently suffering from and, more specifically, about certain illnesses or injuries that may have had a direct impact on how I was able to perform the job I was about to undertake *I was not specifically questioned about psychological issues or illnesses I had, or was presently suffering from –* although there was the caveat in the medical questionnaire that if I was not fully truthful in answering the questions and later it was found that I had concealed what should have been revealed then it would result in the immediate termination of contract, there was no targeted screening of my mental health background.

This is a real blind spot in the current average recruitment policy for CW's; as so much in care takes place on the psychological level then a proper assessment of the background and fitness of CW's' in this respect would seem to be not only a proper safeguard but an absolute necessity, the termination of employment sanction that should it be proved later that an individual had failed to declare in the medical questionnaire what they were implicitly bound to is clearly inadequate if the revelation only follows injury or harm that comes to a vulnerable person because of a carers' psychological illness.

What is being cried out for here – although it is something of a silent scream – is a proper check of the background of CW's that looks at their health history holistically, taking into account not just physical conditions but mental ones too – in fact questions surrounding the psychological should be given as much prominence in a boilerplate medical questionnaire as those of, say, previous back injuries or of cardiovascular illnesses.

The immediate challenge to this suggestion is of course the consideration of how this policy should be employed and policed that raises issues of privacy, confidentiality, the privileged nature of the relationship between doctor and patient and an individual's rights to be seen – ironically, as noted above in relation to service users – to be seen as a whole person rather than through a narrow prism of their afflictions; the only truly reliable way that such checks could be carried out would be for potential employees in the Care Sector to be compelled to give consent to the accessing of certain parts of their medical records. This though raises more issues than it answers – from the obvious - How can potential CW's be protected against discrimination on purely mental health grounds? How long should consideration of past psychological history have relevance to applications? Would an individual's history of self-harm as a teenager say, affect their job prospects as a 20 or 30 something? – to the pragmatic - Who would have access to the information shown in medical records – the personnel department? The administrator? The service manager? - to the legally required storage and retention of personal details and data - How could such information now loosed in the workplace be protected from wider dissemination?

These are difficult issues, and not just in the terms of those covered above but also as applied to a deeper more esoteric level than prohibitions and exceptions. For example my own mental health history allowed me to have a deeper understanding of, and empathy with, many of the people I was to go on to support and care for - having been in the depths of mental turmoil I could identify clearly with those permanently stuck in it. Further, although I may have had my own issues I remain certain that they almost never had an impact on how I did my job or in how I engaged with it. Had screening been better, as proposed above, then I would almost certainly have been barred from care, yet I was a decent and competent CA, at least the equal of those I worked alongside, so why should I have been locked out of a job that I was more than capable of doing and frequently doing well?

The problem in making prohibitions is that there are always exceptions to the rule, and in the case of care there is the very real risk that you will prevent many good CW's from entering the field who in fact go on to bring a deeper insight into the requirements and needs of service users that makes them better carers than those without such mental issues. More importantly it also fails to account for the shortcomings that would not be covered by such a stringent medical/psychological screening process.

For example in my first couple of weeks as a CA I worked with one carer who had severe literacy issues and an almost total lack of numeracy, another who could not keep his eyes open for half an hour straight when he was meant to be supporting a client who was flagged as having unpredictable mood swings that could rapidly escalate into very challenging behaviour, and yet another who was so disengaged from the caring process that it could be said that they didn't actually care at all. All these issues were known to at least the line managers and sometimes to senior managers, as between these three CW's one could not keep accurate and timely records, another had in the past placed a client at serious risk of injury in a potentially major incident and a third had previously been investigated for neglect. Skirting the issue, for the moment, of whether these carers should have been still allowed to be working in care, how reasonable can it be that someone like myself should potentially be excluded from care because of a history of alcoholism and depression and yet these CW's continue in care? Once more – like everything else in care - what appears to be straightforward is actually much more nuanced and ambiguous – yes it may be an excellent idea to screen potential CW's more rigorously but at what cost to carers who may be able to bring another dimension to the support of service users?

What is highlighted in the above is the need for a proper registration, oversight and advocacy body for CW's along the lines in which nurses and doctors are registered and represented.

The Kingsmill Review: Taking Care – An Independent Report Into Working Conditions In The Care Sector (hereafter referred to simply as The Kingsmill Review) led by Baroness Denise Kingsmill CBE makes this very point –

"Care Workers need a stronger voice and greater status. The people who perform this essential job must receive fair compensation for their efforts, be able to input into practices and policies that affect them most, and have their rights protected. Care Workers are paid from public funds, whether through an independent Service Provider or directly by the Local Authority, and it is essential that taxpayers know where this money is being spent. The development of a registration system for Care Workers is an effective way of regulating standards in care and protecting workers' and users' rights. The registering body would issue a Licence to Practice, which gives credibility to the profession, promotes the status of the workers and provides important benefits to the members, while also protecting the public from unqualified, incompetent or unfit practitioners. The Patients Association and other groups called for formal registration to ensure "appropriate feedback and a consistency in recruitment, training and professional development".

The benefits are numerous: protection and legal representation, advice and support for workers, the ability to share best practice, network, a greater sense of status and a symbol of a professional workforce."

http://www.yourbritain.org.uk/uploads/editor/files/
The Kingsmill Review - Taking Care - Final 2.pdf

[The Kingsmill Review goes so far as suggesting suitable registration body – The Health and Care Professions Council – HCPC]

Given the common sense arguments put forward by Baroness Kingsmill it seems incredible that such an unlicensed, unregistered and fundamentally disorganized workforce as CW's delivers some of the most important parts of care to the most vulnerable in our society. It is as if airline pilots or Heavy Goods Vehicle (HGV) drivers were allowed to take to the skies or roads respectively with only scanty training and oversight and even this dependent totally on an employer discretion.

[N.B. At first flush this analogy of CW's with airline pilots and HGV drivers may seem overblown but will be returned to later to illustrate that it stands scrutiny and is not then mere hyperbole]

As things currently stand at there are a number of major conflicts of interests in the recruitment, training oversight and standard setting of and for CW's, the most major of which is the leaving of training, oversight and the standard setting to employers (and obviously –indirectly - the care sector watchdog – The Care Quality Commission - CQC – who's' manifest shortcomings are worse than can be imagined by the public – a topic I shall cover in much more depth later).

Recruitment and training costs money for service providers - The Kingsmill Review states the cost at £3500 to train a CA properly - and so they are averse to firing poor carers, this leaves many CW's who have glaring deficiencies in post at a cost to the service, other CW's, and most importantly, the service users.

The obverse of this point is that service providers – in a sector were turnover rates can reach up to 30% a year - have a implicit incentive *not to train CW's properly* so they can be disposed of more easily, this leads to a 2 tier workforce - those trained to a decent standard whom service providers are reluctant to sack no matter how low their performance falls, and barely trained CW's, who may be very good but who can be shown the door almost peremptorily (we will return to the subject of training in more detail below – here I only want to draw attention to inherent conflicts of interests)

This 2 tier system in turn shines a light on another key conflict of interest in care - there needs to be a better way of monitoring performance and driving up standards than simply keeping it all "in house" - which almost militates for a culture of cover-up and denial.

As well as providing oversight and standard setting a registering body should also act as an advocacy body that is able to defend CW's against unfair, discriminatory or arbitrary allegations and/or disciplinary procedures. At present it is left to employers to decide if and when to punish infractions or deficiencies, this can easily lead to witch-hunts and victimisation, CW's can easily find themselves investigated on often the most flimsy of evidence - and sometimes even this evidence can be manipulated by certain individuals in the care environment - and it is possible for an unfortunate CW's to find themselves sacked on little more than hearsay, as will be shown in more detail in the following chapters.

All these issues surrounding recruitment of carers could to an extent be mitigated if induction procedures in care providers were rigorous and adhered to. It hardly needs saying, after all I have said so far about care, that this is not the case.

Arriving on my first day in work in a job that I had never done before, in a specialised field in which I had never laboured and for which I had no qualifications and only the thinnest of experience of, I was expecting a day of induction, orientation, perhaps a tutorial in the basics of supporting service users and at least a cursory run through of how I was carry out the work for which I had been employed. Instead I was taken into the administration office, had my photo taken for use in my personnel file and for my ID badge – which was so flimsy and amateurishly designed (basically a double laminated piece of paper with a passport sized photo of myself stuck to it) that it fell apart within a couple of weeks - brusquely walked through the company intranet registration and navigation, given a fat wad of papers that were to be my learning file for the basic vocational qualification in adult care – the Learning Disabilities Qualification (LDQ although, as is the nature of care qualifications, this has now probably been replaced) which I was bound by the terms and conditions in my contract to complete before the end of my probationary period - fixed at 3 months - apprised of the company sick policy, asked to sign multiple pieces of paperwork, among which was my contract, and finally I was introduced to 2 huge lever arch files that contained the company's Policies and Procedures - these I was supposed to read by the end of my probationary period - and that was it.

This "induction" - which lasted all of about 90 minutes - was supposed to at least prepare me for work, yet I had not been introduced to my line manager, my Team Leader or any of the care staff I was going to working with nor had I been introduced to any of the service users that I was to support or had any idea of where I would be working.

After completing this "induction" I did finally meet my Team Leader and taken to a cottage where the two service users I would be primarily supporting were housed – this was the first indication or introduction to where and with whom I would be supporting.

Needless to say by this time I was so utterly bamboozled by the speed and brevity of the whole process of "induction" a very large part of me suddenly wanted to cut and run there and then. My naivety in expecting at least a day of basic training and a polite introduction to those I would be supporting and working with was bluntly exposed. And yet, for all its shortcomings on getting to the cottage I was presented with two more huge files, these contained the personal details of the two clients – their history, their medical and psychological backgrounds and their current care plans – all of which – apart from the care plans - were unsurprisingly fragmented as their care had been similarly disjointed, but they were all I had to go on to discover about the people I was to be supporting. At last I was vaguely introduced to both service users I would be supporting then parked in a chair in the living room and commenced to wade through the two huge personal files.

Such was the totality of my induction, yet when I say that *this was the fullest induction I had to any job in care* it is neither a joke nor an ironic comment, it is just a simple fact. As soon as my CRB clearance came through I was expected to support either of these two service users by myself armed only with the knowledge gleaned by my reading of their respective files and by my observation while I was "shadowing" current CW's doing the job; as one of these clients in fact posed a serious risk to themselves and to certain sections of the public then had my CRB come through quicker than it did I could have ended up supporting this service user alone and therefore tasked not just with protecting their safety, security and wellbeing but that of other vulnerable groups of the public as well – a very heavy and serious responsibility to charge to an untried, untrained and unskilled CA.

To say that this posed a grave risk to the service user, myself and others is a colossal understatement. In respect of the time it took for my CRB clearance to come through I was lucky; because of my chequered past work history and a series of "geographicals" during it - in which I had had dozens of addresses and jobs in many different locations across the country ranging from Kent to Scottish lowlands - my CRB took several weeks to clear, during this time I gained vital and invaluable experience and insight into the service users I was to support, however in the case of other people that had been recruited at the same time as me – one of whom was a fresh school leaver – their clearances took less than a fortnight, after which they were left with sometimes very challenging service users in remote locations. As it turned out the fresh school leaver on her 3rd day of going solo was attacked by a service user and sustained not insignificant injuries. She left care all together shortly afterwards to work for a supermarket chain.

The lack of proper inductions in care breaks a link in competencies of care delivery for a service user – they have to adapt to very new and inexperienced CW's who may not be as alive to their needs and requirements as more experienced staff that they have become used to - and provides another dog that doesn't bark in oversight of CW's; what should be a much more rigorous procedure that fully prepares new – and often very young - CW's in the skills, management and responsibilities of care work - is instead tuned into a wholly administrative process that has nothing to do with actual care at all.

The Kingsmill Review focussed on just this point, saying –

"Many new Care Workers receive only 1-2 days of shadowing before they are expected to work unsupervised."

This creates a double hazard given the slack method in which care workers are recruited. The lack of rigour and proper screening in recruitment could be largely ameliorated if CW's were given even mildly adequate inductions that were tailored to the service users that they were going on to support; but as this is lacking it is possible for someone to literally walk off the street and be caring for a vulnerable adult - who may pose risks to those inside and/or outside the care environment, or be put at risk themselves from an inadequately trained CA - within two weeks of employment.

Of course meaningful recruitment practices and a rigorous induction poses one great obstacle for care providers, regulatory agencies – specifically the CQC - and ultimately the government – all this would cost money, planning and careful consideration, and it is one of the many dark secrets in care that the cost of implementing such a clear and interconnected process does not meet a cost benefit analysis. It is far easier and cheaper for government regulators and care providers to keep the cost of care down by maintaining the status quo as the costs of inquiries, compensation and legal proceedings when things go FUBAR are variable costs, whilst those of reforming care and making sure it is policed effectively would be a fixed and structural cost.

Every party in the adult care process - from care providers through to regulators (specifically the CQC but also the Adult Safeguarding Boards of Local Authorities - LA's) and ultimately the government (of whatever stripe) have a vested interest in keeping such fixed costs down – for a care provider it means more revenue - and profit if it is a for profit provider – as barely trained personnel are more easily and quickly replaceable than well trained and experienced ones; for a regulator it means less work for an already overstretched body; and for the government it means that the cost of care is maintained artificially low *in the respect of the recruitment and employment costs* of CW's.

[NB - It should be noted here that evidence of a mentored and written induction of staff is a CQC core standard but in my experience this is only ever rarely checked upon, almost never meaningfully enforced, and amounts to little better than a paper exercise – in other words they look only for an absence of evidence not evidence of an absence).

This whole rigged and ramshackle process manifestly takes no account of the human cost of continuing as we are - for service users it means they are often poorly supported by a badly prepared and trained workforce, and for CW's it means they have to shoulder the burden or carry the can depending mostly on luck.

In fact CW's come off the worst from the system as it currently stands, as being poorly trained and wholly underprepared if they meet with injury or stress related illnesses through being thrown into situations they are totally overwhelmed by then they can easily be replaced; if they provide poor, defective or even abusive care then they make easy scapegoats, either way they are left in no doubt that the buck stops with them.

In that first job it has to be said I was lucky in that although the induction process left almost everything to be desired, once in post I was more than adequately supported both by management and some CW's (but not all, in fact the opposite was true of others who acted toward me and other new support workers in what could only be described as a disgustingly contemptuous fashion - but more on this in the next chapter). I was also given decent guidance in how I did my job.

Yet some holes in the management of myself and other CW's still remained glaring. Although I started as soon as possible on my LDQ and finished it well before my probation time was competed this was not the case with most other support workers, who either never started the qualification or ran into insurmountable issues in tackling it.

Among the first group – those who never even started the LDQ – were those who had no intention of doing it – these were mostly what I would call "stop-gap CW's" that is individuals who have gone into care not as a career choice but because they had little or no alternative to it. These were not just those referred to in Chapter 1 who found themselves corralled into care but rather those who considered care to be a temporary job until something better came along.

In my experience a more than insignificant minority of CW's fitted this description – while the obvious undesirability of having a large chunk of the workforce so minded is obvious, a side effect is that they see no use in investing time and energy in pursuing a qualification in a field that they have no intention of staying in.

This though is not the point, the LDQ is designed to be a basic care qualification and, to the credit of its designers and administrators, it is meant to – and does - reinforce what should be part of every induction – creating an understanding and professional attitude toward those with learning disabilities and making carers aware of the duties they have to fulfil to comply with the law – so to have a section of the workforce not engaged enough to *want to know* these basics and in fact are *resistant to even starting to understand them* says much about the potential quality of care they provide.

Among the others who had never completed the LDQ were those who simply on an academic level could not complete the qualification without full time guidance that was simply not possible to be provided for in a work based vocational qualification. These were individuals with serious literacy issues who found even yes or no questions impossible to answer. This in itself was not a barrier as those with literacy issues were accommodated by being able to answer questions verbally to an assessor, but even this did not account for some of the learning issues they had; more than once I found individuals unable to answer questions because of an inability to *comprehend the question*, even when put verbally. As I have already stated in Chapter 1 it is not necessary that CW's should be dormant rocket scientists (lord knows my incompetence with maths) but the fact that there are those in the care workforce who are unable to tackle the *most basic of qualifications* even when it is put in its *most basic form* is concerning. Whilst I would never say that some people *should not be in care* sooner or later someone will have to grasp the nettle and decide that there must be some baseline that all CW's should be able to achieve; the alternative is totally compromised care. The irony here, as noted in Chapter 1, barely needs pointing out - we have created and tolerate a care system that allows people to become carers for those with Learning Disabilities who themselves have serious learning issues.

Regardless though of whether individuals were unwilling or unable to complete their LDQ one common fact united them - the sanction in the contract that the probationary period demanded the conclusion of the LDQ, and that the failure to do so would trigger a termination of employment - was never enforced; there were CW's working for this care provider for over a year who had only made the barest and most half-hearted of starts on the LDQ and seemed under no duress to go any further with it. This failure was not exclusive to this care provider either; it is so widespread as to be endemic. The CQC will only check on the fact that the LDQ qualification is required in the contract, and, when they do check on individual progress, a CA and employer need only show that they have *begun* to work on it for them to be satisfied, the failure to conclude it can be, and always is, finessed.

This fact – that the most basic of qualifications in care, even when contractually required and backed to by (notional) regulatory enforcement – is almost entirely ignored will most likely will be met by a shrug of the shoulders by most readers or in fact be met with actual resistance similar to that of the recalcitrant CW's themselves – the point, most reasonable people would think, is not what bits of paper a CA accumulates to prove their proficiencies in caring but rather *are they any good at actually caring itself*, and, if they are, so what if they don't possess the basic qualification?

Again there is some sense in this – in fact more sense than most care managers and trainers would like to admit - and is a point we shall return to time and again - as in my personal experience the best carers - the most conscientious, effective, compassionate and responsive of CW's - were those without, or with only the bare minimum, of qualifications. The corollary of this is that the most qualified of carers I worked with were nearly always either stunningly incompetent, thoroughly lazy, or overbearingly arrogant; and frequently a combination of all three. In many respects paper qualification bore no relation to the quality of care staff or, where it did, they were in inverse proportion.

However much sense there is in the fact that care staff should be judged on their caring capacities not on how many qualifications they have sticking out their arse, it misses a vital point – a carer needs to know, and needs to demonstrate that they know, what is required of them by law and by regulation as it applies to the workplace, they need to be aware of what standard of care is expected of them and how to achieve those standards of care – without this when things go wrong a carer can plausibly respond that any failings in their care of service users was because they were not instructed in such matters in which they failed.

The Review of Healthcare Assistants and Support Workers in NHS and Social Care conducted by Camilla Cavendish (otherwise known – and referred to from here on - as the Cavendish Review or Report) saw this apparent contradiction; in looking at the training of the care workforce, she noted –

"But overall, training is neither sufficiently consistent, nor sufficiently well supervised, to guarantee the safety of all patients and users in health and social care. In domiciliary care, we have heard of instances of staff being sent unsupervised into clients' homes with no training......A rigorous quality assurance mechanism is urgently needed, to police a market in which...........too many workers are signed off as 'competent' without necessarily being so."

Yet she also acknowledged –

"Training should not be seen as an end in itself: what matters is that workers are competent, and kind."

Her solution – a "Certificate of Fundamental Care" while well intentioned and aimed at tackling the issues highlighted above, raises as many issues as it does answers, and indeed falls into the gap that exists now between what is a good carer "on the ground" and what *looks like* a good carer *on paper*. Although this is an issue we will return to later it should be said at this juncture that what is sorely lacking is not just training of new staff but the whole recruitment process and thereon the status that CW's currently hold – when recruitment is based on cost and availability and CW's are regarded by employers and the public as so many disposable biological units of labour, no amount of training will produce a caring and motivated workforce.

If these gaps exist in an environment where management is engaged and relatively strong what happens to inductions where management is weak or non-existent?

When I went into care of the elderly I found out.

In my first day in my job at a nursing home I was taken straight from reception up to the EMI (Elderly Mentally Ill) floor where I would be working that day, armed only with a sheaf of papers that were to be my basic induction.

Nothing can prepare the unprepared or initiate the uninitiated for EMI nursing in the private for-profit sector, here the most challenging and mentally ruined of elderly individuals are "cared for". The flaws and shortcomings of such an environment will be dealt with later, but the visceral attack of first impressions are sure to - as indeed they did on me - make a deep and lasting impression.

The sound was first assault on the senses – there were residents in this EMI unit who constantly yelled or shouted or screamed to such an extent that it felt like you had been transported back into Victorian Bedlam. Next was the smell, the steep pungent urea fumes that seemed to stick to you like glue and great wafts of the most noxious smelling faeces odours that clutched not so much at the nose as at the back of the throat. Then there was the constant stream of images - the dead eyed wanderers, those sat in a state of half sleep/half narcotised bare conscious waking, those clothed in food or excreta soiled clothing, others - women and men - in states of half undress, and still others pulling at your clothing demanding attention - for what they were not sure.......only it was *something*.

Surveying this scene I have to admit total shellshock overtook me, everywhere I looked had an image of something out of Dante's upper circles of hell rather than a care environment.

I defy anyone on first experience of a busy EMI unit to fail to be anything other than half terrified by all these aspects of suffering that fall upon them.

How much more important then that a decent induction process should be started as soon as possible, and by a qualified and relevant member of staff.

How much more damning then that such an induction was not forthcoming.

I was supposed to have been taken through my "induction" by the senior carer on duty that day, only they were off sick, so the job of "inducting" me was kicked back to the next most qualified CA who appeared vastly underwhelmed at the prospect of having to spend time with this greenhorn; further, this being the start of the shift, which was 12 hours long – 8am to 8pm - there was simply no time to even contemplate beginning it, instead – and having not been in the building more than 20 minutes (and yet to have my CRB clearance come through) - I was asked to change a continence pad on a heavily soiled resident who was particularly challenging, to wash and dress them and to do it completely on my own as the other staff were "busy". Without knowing anything about this resident or, more importantly - considering I was being asked to perform a particularly intimate task - without the resident having a clue who I was - I was left to struggle on alone with a task that should have involved 2 carers. Loath to question and eager to impress I assented without demur and battled on solo in a welter of effluvia with no assistance being available at any further point or anyone checking that what I was doing was right. I eventually got the resident clean and changed but only after a good 20 minutes of sweat and toil and no little discomfort for a deeply troubled service user.

Then it was breakfast - which seemed more akin to a *Sturm und Drang* full-on combat firefight – only with nicer decor - than a meal; residents wandered in and out of the dining room others were left to eat with their hands, one particular resident even bypassed this rudimentary method by gently tipping up their plate and putting their face in the food and taking mouthfuls between coming up for air; to put it lightly then, it was clear that several of the residents needed assistance with eating; but while CA coped as best they could with the majority of these, two others stood idly by talking, drinking coffee and doing a brilliant job of blithely ignoring the chaos around them, thus leaving at least 2 residents who needed support in feeding themselves left alone staring uncomprehendingly at their plates of rapidly cooling food.

[N.B. Although we will return to this point later the insouciance of these two carers was not an outré example of staff "quality" in residential care for the elderly – nearly everywhere I looked, on every unit in the nursing home, I saw in the majority a total, determined and committed dedication to work avoidance – this was my introduction to what became almost an observable and iron law in care of the elderly, what I would call the 90/20 rule – in any care environment 20% of the workforce deliver 90% of the care]

To say the whole process lacked civility would be an abuse of language, to say dignity, care, compassion, or respect were lacking here is akin to saying a black hole is pretty big and quite dangerous.

However I didn't have much time to stand there – chin on the floor with horror, totally stunned and rooted to the spot by the Hieronymus Bosch-esque tableau before me - or to support either of the two neglected, unfortunate residents blankly contemplating their congealing bacon and eggs - as a feeding cup and bowl of pureed sausage, tomato and bacon were jammed into my hands and I was directed to the room of a very infirm resident in the latter stages of dementia who was bedbound and completely unable to feed or hydrate themselves and told with an airy wave by one of the idle CW's –

"Give that to [name withheld]"

This resident was particularly difficult to feed and whole process of feeding and hydrating them took at least 20 minutes, when I emerged from this residents' room all the breakfasts had been cleared away apart from those of the 2 unaided residents – who were still staring at their – now - stone cold breakfasts and unsurprisingly resisting being fed any of it by the one diligent CA; the others were nowhere to be seen.

Such was my main introduction to care of the elderly.

Later that morning I had my "induction".

Snatching time in the hiatus between breakfast and lunch the reluctantly appointed carer raced through the induction booklet with dizzying speed signing off on as many processes as possible. What was more worrying than the speed though was that this carer – who was the most qualified among the staff on duty and the most experienced - *did not themselves even know* some of the policies and terms referred to in the induction booklet. Section after section, totalling at least a 3rd of the booklet were passed over with the repetitive phrase of –

"You'll have to ask X about that [referring to the absent Senior Carer] when they get back."

The whole process seemed to me to about getting a totally administrative exercise done as quickly as possible so that my presence could be utilized as another pair of hands. This was all far from unusual as regards to inductions with that particular care company, in fact it was to become the norm to me the longer that I worked in care for elderly and my experience of other care providers.

It so happened that I was somewhat lucky – (this is what passes for luck in care) although I was stunned by the visceral nature of care environment I found myself in, I had a least a background in dementia care (see Chapter Moving On) and although I wasn't prepared for the intensity or the reduced level of care – due to that intensity - I was asked to provide I was able to cope with the fundamentals. More than that though was the fact that I had initiative and willingness and so was quickly accepted as a decent worker because I was ready from day one, minute one, to get "stuck in".

I was though different from a lot of new starters. As pointed out in Chapter 1 many new starters are new to care or even new to the very world of work, to have them thrust into such difficult and potentially traumatic work atmosphere and not to give them at least a basic grounding in responsive, compassionate and skilled care provision, as well as to orientate them in the infinite complexities of working with dementia sufferers or those elderly with serious and often terminal medical issues - never mind a background in company policies and requirements - is to ask the moon and stars of immature, uncertain and often less than motivated individuals.

Lacking the foundation of a proper induction, confused, apprehensive and uncertain, a new recruit often finds themselves – like myself - thrown in at the deep end - and very deep it is too - and because the care workforce is busy they will have less or no time for showing a new recruit how any work is - or should be - carried out; in fact the very time out needed to educate someone new in the ways of work will often be regarded as an inconvenience or distraction; if the recruit is new to the field of care this only makes things worse as the very basics have to be gone over. Very often in such instances case it is easier for an experienced worker to simply do the job themselves – this creates a vicious circle - a new recruit, knowing little, will be given less chance to actually find out how to do certain tasks and when they attempt them they may get them wrong so creating even more work – a job done badly that needs to be unpicked is often worse than a job not done at all - and so reinforces the practice of excluding a new worker form further tasks.

Due to this process new workers will often be labelled as lazy from the very start simply because they do not know what tasks need to carried out, how they are to be carried out and when they need to be carried out. This creates a totally dysfunctional system, and this is what every care environment I have worked in is like; dumped in a strange and alien workplace with people too harassed, busy or belaboured to take the time to get to know them never mind teach them, new CW's will often be left floundering. The results of this are always baleful – they are why accidents and incidents happen and they breed an endemic culture of isolation – both for service users and CW's – in a field where teamwork is not just necessary but vital.

All this would be understandable if there were a benefit to the employer in allowing induction procedures to be short circuited but any such benefits are so painfully short-sighted and short term that it makes you wonder how any care providers actually function on a paying basis seeing that the care of their most precious commodity – carers – are treated in such a slipshod manner (how vital CW's are to a care provider should be easily apparent but more on this shortly).

Often new recruits will be thrown into the work environment simply because they are another "body" – that is their presence has been counted by management as part of regular staffing levels - in order to save time or money; as such new recruits are replacing experienced CW's this means there is no chance of a rookie learning anything. While this may have a benefit in the immediate term, in the medium and long term it is obvious how self-defeating this is – poorly inducted staff will not know the correct way to do any tasks, the time that they have to do them and where they have to do them, this not only increases the scope for accidents and neglect (never mind the compromise of dignity and respect) but also critically – from the employers point of view - lowers productivity – a CA who doesn't know what task they should be fulfilling at any given point in time will be less productive – throwing another body into the breach without proper induction will be just that – throwing another body into the breach, not throwing a safe, effective, caring and responsive body into the breach.

For a new CA the effects can be brutal. Care, by the very definition, means caring for individuals who have challenges and needs that cannot be met in a less intensive environment, this means the support of them is testing mentally, physically and psychologically – this takes all of the carers attention nearly all of their time, if a CA has not been properly trained or inducted there is no chance of them meeting and overcoming the challenges that care poses on a physical or mental plain, in such circumstances it is easy to be overwhelmed.

Many CW's remain damaged by their first introduction to care – they either are put off care for life – I have seen many young CW's last less than a morning and are never seen again – or they adapt in ways that are less than optimal. However it is *how a new recruit learns to do their job* rather than *what they learn* that is of primary importance here; without the support of an engaged management structure and a proper induction procedure new recruits will be thrown back on to learning on the job and it is here that the dice is rolled on both care practices and the adoption of them, as it depends *who* the new recruit *learns from* as to determining *the kind of care worker they become.*

If they are fortune they will be working alongside good, experienced and knowledgeable CW's who can pass on excellent ways of working and methods of support, but, sorry to say, the chances of this are less than good; instead they often will be dumped with CW's who may actually care very little, carry out tasks in a poor or dangerous manner and fail in many aspects of their duty to support those they care for.

More than once I have seen decent individuals become very poor CW's because they have learnt from other more experienced staff whose work and care – or lack of it - leaves a lot to be desired. The fact that most CAs have to learn most of their job while engaged in doing it can hardly breed confidence in those who have entrusted the care of their loved ones to a care provider. Reasonably enough they will expect their loved one to be supported by well trained professionals – how happy would you be to think that the care of your mother, father, brother, aunt, uncle or whoever was being carried out by someone whose total experience was gained by watching others do their job badly? Worse - how happy would you be to know often critical care tasks have been delegated to someone who only started their first job in care 2 or 3 hours ago? No one would dream of letting a 16 year old school-leaver loose at the controls of JCB, or in charge of wiring a house, or fitting a boiler without any proper instruction, training or oversight, only expecting them to teach themselves through watching other JCB drivers, electricians or heating engineers, or to pick up titbits of knowledge if and when such skilled professionals have the time to or inclination to show them; yet the care system allows such inexperienced youngsters to look after the wellbeing of frail and vulnerable individuals as a matter of course.

Perhaps the most damaging result of this process though is that it fosters what has become to be known as a "care worker led environment". This construction and the situation that it denotes is perhaps the most important aspect – and failing - of care as it stands today, and as such its importance cannot be overstated.

CHAPTER V

Wagging The Dog

"It is not clear how the boundaries between the nurses and Support Workers [CW's on my terminology] were developed or maintained, or how the division between their responsibilities was determined. What appears to have happened is that, despite the presence of a team of 13 professional nurses including managers, over time, Winterbourne View Hospital *became a Support Worker led* [my italics] hospital.

South Gloucestershire Adult Safeguarding Board, Serious Case Review.

The above quote from the inquiry into the colossal goat-orgy that was the Winterbourne View scandal makes clear how some of the practices that were so profoundly depicted in the BBC Panorama programme - Undercover Care: The Abuse Exposed - had their genesis.

To many Winterbourne View is considered as an outlier, an aberration of vile acts conducted by almost psychopathically sadistic staff completely divorced from any control, oversight or restraint.

The truth is – and I'll start a new paragraph here to emphasise my point, and, oh what the hell, I'll put it capitals too – EVERY CARE ENVIRONMENT I HAVE WORKED IN IS, TO A GREATER OR LESSER EXTENT, CARE WORKER LED. But what is perhaps even more shocking is that all the conditions that prevailed at Winterbourne View *are in place in every adult care setting that I have ever worked.* The reasons for this are myriad but the actual fact that this obtains comes down to simple arithmetic and observation.

In arithmetical terms, in a service, the most predominant staff in terms of sheer numbers are CW's, they far outnumber managers or nurses who are supposed to be control, and in charge, of CW's, and it is simply not practical or possible for senior staff or managers to supervise every action or task that CW's undertake.

In observational terms the facts are just as obvious – CW's have by far the most contact time with service users and so are in unique position to influence, control and even abuse service users, this means that not only do CW's have advantage in numerical terms they also *exert a power far out of proportion* to their supposed lowly status, it can be said that while managers may direct a service, the manner and the quality of its delivery comes down to largely unsupervised CW's.

This leads to a paradox in care – those who are paid the least, treated the worst and who are infinitely replaceable have by far the most critical impact on the lives of the vulnerable and ill people that a service is at least supposed to support or care for. In other words incentives and status are inverted, instead of those delivering the most critical part of a care service – the actual care – being respected and remunerated in a fashion in accordance with their critical inputs they are at every turn de-motivated and belittled, what this creates is an atmosphere in which, as all key indicators of worth are lacking, CW's will look for other ways in which their very difficult job can be made easier, more interesting, more self-serving and more self-affirming – in short they will produce a care service that is run *by* and *for the benefit of* the CW's rather than the service users.

Another paradox thrown up by this mismatch of responsibilities and value is that although CW's are where the rubber meets the road in care terms they are the most unregulated of all care jobs.

The South Gloucestershire Adult Safeguarding Board, Serious Case Review again –

"What is clear is that professional standards and codes of practice had no bearing on patient care. Although different work groups are subject to differing standards and regulators, the largest group of staff – the Support Workers – were, and remain, without any professional regulation and are not subject to any of code of conduct or minimum training standard."

The fact that CW's were, and *still* remain free of direct regulation (although it will be argued – but not very convincingly – that the CQC has overall oversight of CW's) is nothing short of – literally – criminal; that the most key component and work group in any care environment - CW's - are, in words of the Serious Case Review "without any professional regulation and are not subject to any of code of conduct or minimum training standard" defies rational and reasonable explanation. The absolute need for a professional body that both represents and oversees CW's is as desperate now – perhaps even more so now – than it was when the Serious Case review released its report in 2012.

The fact that CW's are the most vital and the most influential of all the care jobs should be beyond doubt - for all the clear evidence I have mentioned above; but these are merely facts that state what should be obvious. More key, I would argue, are the reasons *why* and *by what means* CW's come to run a service for their own convenience and, sometimes, benefit. These are what I would call "the mechanics" of how CW's come to dominate a service and although they are not as clear cut as the *fact that they almost always do* they are every bit as critical in defining a service at the point of delivery.

Before I come to *why* a service comes to be ruled by care workers responding to their own needs and requirements the *means* need to be illustrated - as in cause and effect - these are profound and disturbing - and became paralysing clear through my experiences on my very first day in my first care job.

Weighed down by the two huge files containing the sum total of the history, medical and psychological backgrounds and the current care plans of the service users I would be supporting I was introduced to the care team who were on duty that day. Both these care workers –one male and one female - had significant (for care) experience in the care field and both could be said to be more than competent in their roles. New as I was I'm not stupid and what was obvious, even on that artificial, stiffly formal, first meeting, was that the male was the alpha dog of the team, that it was him rather than the Team Leader who ran the service I was entering. He didn't partake in any ostentatious imposition of authority – like so many other alpha dogs were to do later in my career – but merely with the way he looked me up and down – literally as well as metaphorically – it was clear that he was "evaluating" me, my worth, my immediate use to him and also - with all the veiled signals appropriately on view – displaying his command of this territory.

Not all alpha dogs (and these can be women – in fact are more likely to be so due to the predominance of females in the care workplace) are so obvious, while some rely on open dominance others rely on more subtle means, but two factors unite them – they are never the person formally in charge of the service – neither the Team Leader nor manager – in fact they nearly always reject positions of authority as they realise that this entails responsibility and an alpha dog needs to avoid this at all costs.

The second uniting factor is that for any new CA it is essential – if they are to survive in care – that they identify the alpha dog and know not to get into any situation of confrontation or disagreement with them. The fact is that while those in formal authority may have the stripes the alpha dog has the bite, any CA crossing an alpha dog is almost certain to have a very short career in that service and perhaps in care itself, I have seen more careers ended by alpha dogs than I have through formal terminations of contract. What was to follow only reinforced this first impression.

I took a seat in the lounge partially out of view of where the two CW's on duty were seated – at a dining table at the rear of the lounge, ostensibly filling out paperwork. Of the two service users one sat in his chair watching daytime TV – for the purposes of anonymity I will call him Service User "A" - and the other – Service User "B" was sat in his ground floor room watching a DVD. Far from interacting with the either client – as the very care plan I was reading at the time instructed that care workers were to do – they both remained at the table. More than this they were talking in clearly audible volume - even though the care plan for the Service User "A" present in the room – because of free floating paranoia – defined this as a serious no-no as it was a potential trigger for challenging behaviour in them (believing that any conversation that excluded them was about them). The Team Leader had left by now to go back to the office to fill in the massive pile of paperwork that his job entailed and, what followed, as I couldn't help but hear, was perhaps the most sustained, vitriolic and shocking display of character assassination and destruction that I had ever heard.

Now I was at that time, although new to care, far from being naïve, I had done more jobs up to that point than people usually manage in a lifetime, I had seen more work bickering and back biting than most and in at least one area of work I had witnessed actual fistfights as a way of resolving personal disputes. What took me aback here was both the length and the depth of the abuse heaped upon other care workers, the Team Leader, the line manager and the charity's management. Over the course of two hours – I eventually timed them, so unremitting was the flow of bile that was issuing forth - they questioned and criticised the mental capacities of fellow care workers – they were either stupid or lazy - the competence and decisions of the Team Leader and his very basis for making them – judging them to be corrupted by self-interest and greed; the mental balance of the line manager – they thought she was insane or at least a psychiatric basket case; and the probity of those running the charity – accusing them of milking the service users. This was far from the usual griping of shop floor workers against management – I was familiar with all of that – this was different in many ways. It was systematic – attacking every element that they perceived as a threat to the way they wanted the service to be run; it was calculated, vicious – in both the insults toward individuals and of their character and motivations; and it was gratuitous - the very indulgence and pleasure they seemed to be taking in the whole process was verging on the wanton.

One of the most disturbing characteristics of this verbal assault was the racial element that it exposed by its very concealment. The Team Leader of the service was Asian and one of the care workers a black African, in their abuse of each of them these two CW's were very careful not to mention the words "nigger" or "paki" (the Team Leader was actually Indian but racists are not known for such "minor" distinctions such as nationality in their abuse) but it was inherent in the very language they were using, the careful nature to dance around the issue of race, the phrasing of how and why the Team Leader was unfit to lead the care staff, and the inference that the black African CA - who was Nigerian – was both lazy and perhaps on the fiddle in some way - were both stereotypical and demeaning. The unspoken nature of the insidious racism was in many ways more disturbing than if they had actually used the taboo words that I mentioned above, because it gave an insight into the very motivations behind some of this abuse. Racism does not exist just in language – in fact a concentration on language as the racial space is distracting – it exists in attitudes, the way people are treated and viewed differently simply because of their ethnic extraction. More will follow in depth about racism in care in a later chapter but here it is important to note that very often in care white poorly paid CW's will frequently find themselves under the direct formal authority of someone from overseas and this, as can be imagined, produces a veritable breeding ground for all manner of racial confrontation.

Although it didn't occur to me at the time it was only later that I realised that while this vitriolic conversation was a common occurrence, and would have taken place whether I was there or not, at least some of it was no doubt carried out for my benefit, but some of it was also was meant for the ears of Service User "A".

For my ears they were both marking my card as to who was inside in what emerged as their small cabal – three CW's – the two then on duty and one other – a small number but in a care staff of 7 (including this cabal) more than enough to totally subvert the service – and attempting to form my first impressions – by running down other care staff and the whole management structure.

For Service User "A" – although they had severe learning disabilities and behavioural issues, they was far from ignorant of the world around them, in fact they were, in many areas, as sharp or sharper than many other individuals I have met in common life – it was intended to mould teir perceptions of the qualities and value of various members of their care team. Hearing the views of these two CW's – who Service User "A" had a good rapport with and trusted – they would no doubt have taken on board what was said and this would have formed and reinforced the impression that the abused care staff were not up to the job.

In the case of both the service users *I know that this had an effect*. In Service User "A" this manifested itself with increased challenging behaviour when the abused care staff were on duty as compared with when any of the "cabal" were working, also in these episodes Service User "A" used much the same language toward care staff that I had heard from the two CW's handing out the abuse on this occasion. The effect on Service User "B" – who manifested no challenging behaviour and who was also very intelligent despite their learning issues – was much more subtle. Service User "B" had a close and supportive family who were still very much engaged in their life, as well as an active Local Authority (LA) Care Manager all of to whom Service User "B" could be relied upon to pass on - selectively - much of what he heard and witnessed; I say selectively because again he had a good rapport with the "cabal" and therefore had every reason to effectively report back on their views.

This raised many troubling issues. First and foremost it should be said that what was being engaged in by the "cabal" was abuse, clear and simple – the knowing manipulation of service user's perceptions to ends beneficial to some care staff was not only cynical and depraved it also denied the rights of the service users to the best quality of care possible and that of them to make their own autonomous choices and decisions – it was using service users for nefarious ends and therefore abuse.

Secondly I would like the reader to ponder the mind-set of those who were so manipulating the service users in this case.

This was not an instance of something being said in the heat of the moment or in a fit of frustration – of which in care there are many – this was a clear eyed and calculated strategy to make life as difficult as possible for certain CW's while enhancing their own positions.

Think what kind of individuals would do this - to plant seeds of disharmony that would only come to fruition over a period of time and only with the aid of constant reinforcement and reiteration of their views; think of how this was effectively programming a vulnerable individual to carry out actions when the "cabal" were not there; think of the knowing cost that this would impose on certain CW's in terms of their self-esteem and career prospects; think of all the ramifications of these actions - and if you have any sense of right or wrong then you will be appalled, like me, that individuals like this should be trusted and valued members of a care team charged with caring for the ill and the weak.

Of course this is not the form of abuse that catches headlines (or indeed that is caught at all) – it involves no physical abuse, no plaintive victims and no explosive footage, indeed if filmed, and out of context, the abuse of other care staff that I witnessed may look, to the untrained eye, as nothing more than griping – however colourful and vitriolic. This though only makes the abuse worse because it hides in plain sight and therefore is almost impossible to eradicate, it also has more than one victim; not only is the primary victim – the service user – abused, and abused without their even realising it - but so are the care staff who have to deal with the consequences, who are on the wrong end of the programmed responses in the service user.

The truly depressing fact was that this strategy was so successful - the programming of the service users became a self-fulfilling reality; Service User "A" was more challenging with the CW's who were being run down which made them look less competent than the "cabal"; and Service User B repeated the criticisms he had heard of other CW's and so doubts were cast by outside care professionals as to the quality of the care they provided. In the case of Service User "B" this factor – among others we shall come to - eventually led to them being removed to another service by another care provider because the family and outside care professionals came to doubt the quality of care they were receiving.

For the "cabal" this strategy had a double benefit, while other care staff saw their stock fall the "cabal" saw theirs rise, their record of having less challenging behaviour episodes on their shifts compared to other CW's made them look much better CW's than they were, also the good reports made by Service User B – unwittingly – to their parents and Care Manager enhanced the reputation of the "cabal" further.

Through these actions the "cabal" had come to be seen as critical members of the care staff and therefore needing to be retained at all costs, this is what effectively handed these CW's control over the whole service, and they used this power with devastating results.

Through the manipulation of Service User "A" they had two other CW's investigated – a horrible and Kafkaesque situation to be in for anyone who has been through it - for neglect, another two CW's were driven from the service because they could not deal with the stress, and the line manager of the service (who I would later cross swords with myself) reduced to a nervous breakdown when Service User B's Care Manager questioned their competence and challenged their every decision. Each of these episodes were carried out to consolidate the "cabal's" power – the two CW's investigated were the only two CW's to stand up to the "cabal" and therefore were considered dangerous to them, the two CW's that were driven out were simply not liked by the "cabal" (they too also "happened" to be Black Africans – strange that don't you think? Irony intended) and the manager who had the temerity to question the quality of the care that the "cabal" were providing – which was inversely proportional to the power they came to wield – was so damaged by the "cabal's" actions that she never again challenged them.

This is *how* CW's produce a "Support Worker led" service - the accumulation of power through the most devious and cynical methods and then the use of that power to produce conditions that *the CW's want to prevail.*

The results are what I witnessed in the service where I worked.

More important than even this though is where a "Support Worker led" workplace can lead. The workings of a "Support Worker led" environment were all on display at Winterbourne View *before* the Panorama investigation, as can be seen from just one of many similar extracts from the South Gloucestershire Adult Safeguarding Board, Serious Case Review remarking on the intransigence of, and control by, CW's in the institution. The following was taken from notes of a nurse's probation review, notification of suspension and investigation into "capability" and date from a full year before Panorama broke the story of mistreatment at the hospital (May 2011) but around the time that reporter Joe Casey was filming his shocking content. A nurse -

"felt [CW's] were surprised that she was instructing them to complete tasks...tried to explain that when she is Nurse In Charge she needs to know what is going on and where everyone is on the ward....... said that she felt there were too many restraints going on upstairs and patients are being restrained for the wrong reasons. The nurse said she felt that staff needed more training and skills in how to verbally de-escalate situations...she finds the staff more difficult than the patients."

The Management response that she was –

"[The nurse was] rude ....negative and unhelpfull [sic] towards staff...[had] not been informing the management at Winterbourne View of serious incidents...[had] not been engaging in the restraint process, even though [she was] fully trained to do so...[was] not following clear instructions…not partaking in handover process… not adhering to observation levels…not meeting deadlines."

With the benefit of hindsight we can view all the nurses comments as prescient, outlining, as they do, an out of control workforce being a law unto themselves – unwilling to take any instruction or notice of senior staff, engaged in unwarranted restraints and being "more difficult than the residents."

But it is the management's response that is far more illuminating in terms of how the tail – the CW's - came to wag the dog – the service. It is easy to read between the lines of the management response to the nurses' complaints to see how a weight of CA resentment lay behind the claims against her. She was alleged to be "rude" and "unhelpfull [sic] toward staff" - a staff though who were resistant to any form of oversight and didn't seem to recognise any form of authority that did not come from within themselves; that she had "not been engaging in the restraint process, even though [she was] fully trained to do so" something that in the light of events does her more credit than demerit, but which also reeks of CW's fearing that her dim view of the constant use of restraints made her dangerous – by her not being complicit in the unbridled (no pun intended) use of restraints the CW's *knew were wrong* could mean her taking her concerns to outside authorities - and that could lead them into all trouble; thus the need for her to be neutralised.

As for the rest of the management's criticisms, it is all so much managerial flannel, used when seeking to bring specious claims against staff to book – "not following clear instructions…not partaking in handover process…not adhering to observation levels…not meeting deadlines" these are all largely subjective evaluations rather than concrete facts and as such can be used to besmirch someone's professional conduct without much as a shred of hard evidence – any professional in care will at one time or another fail to follow clear instructions, just as they may well cock up handovers and observation levels, and as for failing to meet deadlines, I know of very few in care who do manage to do this. This all looks like simply like what is was - the management coming under pressure from a powerful – too powerful – CA workforce to get rid of a senior staff member who was challenging their way of working.

This conclusion from the evidence may be said to be sketchy and my coming to such conclusions seen as overreaching, but for the fact I have seen exactly the same situation evolve - most especially when I moved over to residential care of the elderly. Nurses or senior carers who challenged staff invariably came off worse because, as mentioned above, they are in the minority, lack a powerbase and are totally unsupported by management. At least two very good nurses I worked with (and for the record who I didn't like on a personal level but respected and obeyed professionally) but who had abrasive manners and forceful personalities were eventually shown the door - snowed under by specious staff complaints, intransigence, and, at times, organized refusal to work for them (in one case this involved a mass phoning in sick on several of the shifts where one of the nurses in question worked). In the face of this management, rather than facing the problem - that the workforce were beyond control - merely caved to their demands, and so handed CW's even more power.

All this should show *by what means* CW's come to control a service but one thing already covered needs to be re-emphasised here as it is important to all that follows - although it is not eye-catching, although it is not newsworthy and although it is never reported, it is critical - BY FAR THE GREATEST AMOUNT OF ABUSE IN CARE IS PERPETREATED BY CW'S ON OTHER CW'S.

This could be commonly described as bullying, yet in care it goes beyond that. Because care deals with vulnerable, ill or infirm service users, and outcomes and performance indicators are measured by the effects on those same service users, the capacity for some carers to manipulate situations and service users to their own advantage are legion. Also because service users often lack sophisticated communication skills this process can go on for some time with no real evidence of it happening and this produces and added twist - carers can seem to be targeted by certain service users rather than their real tormentors – other CW's. In the highly pressurised and complex field of care this can drive carers to the edge very quickly and produce the most undesirable and unforeseen results. It is my opinion, and my experience, that by far the biggest factor in the high turnover of staff in care is down to abuse handed out by carers toward other carers.

This all begs the obvious question – why. The answer can be largely summed up in one word – boredom.

Although care is often understaffed, hectic, stressful and rewarding it also involves long periods where the mind can remain disengaged from any useful occupation. This shouldn't be the case because in care – despite the largely lightly educated nature of the workforce – constant thinking and engagement are vital to good care; not only does it help to anticipate difficult or challenging situations and so prevent them before they start, it also keeps the focus on the service user – what are their needs? What are their requirements? What are they telling you (in verbal or non-verbal communication)? How are you to interpret what they are telling you? How do you react to what they are telling you? – All these things often need to be considered upon an instant and this situation is likely to persist through the majority of your shift.

Of course it is impossible to be this engaged through 12 hours so everyone, to a lesser or greater extent, disengages from the process at some time or another. Some CW's though – and I would class a large minority in this, if not a majority – *never fully engage in the first place*. These are often poorly trained and de-motivated carers or, and let's be honest here, the downright lazy. These CW's have, because of their disengagement, lots of idle time on their minds for long periods at work and this breeds boredom; without the stimulation of wanting or needing to do a decent job they are left to ponder other things. Tolstoy wrote that boredom breeds viciousness and I can attest to that fact by witnessing it; with little useful mental occupation many CW's invariably move toward what may begin as petty disputes but which can escalate into the most heinous actions.

On that first morning of my first day on the job what I was witnessing was the symptom of prolonged boredom, the two service users did not have huge personal care needs nor did they have complex medical requirements beyond simple medication, as such they were largely independent, the CW's were there to support them to do a handful of tasks they could not do unsupported. This left a lot of thinking time with nothing much to stimulate the CW's, in such an environment all the worst that was in these people came out.

Once more back to Gloucestershire Adult Safeguarding Board, Serious Case Review - which itself highlighted on Page 2 of its report how the "ways in which staff boredom were expressed" led to abuse –

"The under-occupation and boredom of.........[the]staff was striking. A woman support worker with six years' experience was filmed "casually poking" the eyes of [a service user]"

This is obviously an extreme example, but make no mistake, the viciousness, calculation and intent to harm exists in almost all care workplaces bubbling just below the service; the fact that this manifests itself in CA on CA abuse hardly lessens the underlying motivation behind it.

However much boredom does play a role there is one other circumstance that, although subsidiary, need to be taken into account; and that is the often chaotic personal circumstances of CW's themselves. No starker illustration of this comes again from Gloucestershire Adult Safeguarding Board, Serious Case Review into Winterbourne View; fully two years before the Panorama programme was broadcast, in 2009, it was reported that -

"Disciplinary proceedings concerning a member of staff commenced. The staff member had been suspended because on the night shift, 12 – 13 September 2009, she had left the hospital at 22.10, and did not return until 12.15, without notifying the nurse. She had had a tattoo done in the car park. In a letter (undated) to the manager she stated, - 'I had no fixed address to return to when my shift finished. I have been experience [sic] severe family problems'."

Hard as it may be to believe, this sort of dysfunction in care staff is not uncommon, although often much less severe and spectacular than in the above case. As outlined in Chapter I care attracts a disproportionate number of the damaged to its ranks – people who have suffered trauma in earlier life or who are escaping the preasent chaos in their own lives by exercising a degree of control over the lives of others.

I have worked with those who have been abused as children, beaten or abandoned by husbands, been brought up themselves in care or by distant relatives, who have children who are beyond all control, who have addiction issues with drugs and alcohol, and, yes, who are even effectively homeless.

Even with myself I could look at my own background and wonder if somehow there was something in care that attracted me just the same as these ill-assorted fellow workers. Of course it would be ridiculous, criminally intrusive and clearly impossible to insist that all care workers should meet the criteria of some absurd "normative" domestic circumstances, however Care Providers – by that I mean employers - more than any other employer I have worked for personally, or have had second hand experience of – *are the least interested in the overall welfare of their staff.* In any care company I have worked for – large or small - I have never even encountered or heard of an occupational health team or of any mechanism of support for workers beyond what they could get from managers of the service they work in. Add this to an endemically dysfunctional workforce and you have a recipe for trouble.

The question that always hangs above care and which is repeated over and over in my experiences is how is it that those charged with looking after some of the most vulnerable in our society are so poorly vetted, so inadequately supported and allowed to be so little occupied? The answer comes from a surprising quarter – the Financial Times - which in 2011 - in the aftermath of the Collapse of the Southern Cross group of care homes - carried out an in-depth investigation of care of the mentally ill and old in residential settings, its conclusion was the "most brutal" fact that –

"However much we may wish to think otherwise......Britain collectively does not care that much about its elderly, handicapped and mentally ill.....The private health sector employ[s] mainly lowly qualified and lowly paid staff who care, almost by definition, for the less articulate."
http://www.ft.com/cms/s/0/5b4fcbc0-8ba1-11e0-a725-00144feab49a.html#axzz1NkYuFORY

The combination of low standards, low pay and poor conditions and a client base who are little able to make representations for themselves as to the quality of their care is almost guaranteed to produce perverse results; at the extreme this produces either chronic lapses in care as witnessed at Stafford Hospital or extremes of brutality evidenced at Winterbourne View.

In between is the sort of bullying, intimidation, isolation and de-motivation of care staff by other care staff, and, most damagingly and worryingly, the psychological abuse of service users for CW's own cynical ends such as that which I witnessed on my very first day. The result of all this is that the care sector that provides for the mentally ill and elderly is endemically dysfunctional.

As evidenced above, and most depressingly, the system as it stands rewards poor or manipulative care staff and destroys or moves on good CW's; however as the dysfunction in care is endemic, good care staff will always run up against the same issues no matter how often they change jobs; this and this alone is why care has such a hugely wasteful turnover of staff.

All of this is pretty much hiding in plain sight, the public are aware that care staff are often paid minimum wage, work long and draining shifts and are daily overstretched yet do they agitate for better pay and/or conditions for CW's? If anyone has heard of massed rallies in support of better pay and conditions for CW's then they have heard more than me. The truth is the public are only too complicit in this state of affairs because, at the end of day, it keeps care costs down – given the choice between the current system and being asked to pay another penny on then pound in taxation for care that is worthy of the name most people appear to settle for the status quo.

Yet maybe, just maybe the public would have more sympathy, and be willing to pay accordingly, if they were aware of the real responsibility that is placed often on the narrow young shoulders of care staff; this responsibility, and its consequences, are massive and yet many remain unaware. This is not the public's fault, even after several weeks "shadowing" care staff I was unaware of the weight of responsibility, it only hit me on the first day when I was left alone with a client who's quality of day depended on my quality of care for them.

CHAPTER VI

Reality Check

There is a psychological phenomenon called The Illusion Of Control that – to quote Wikipedia - itself following Suzanne Thompson -

"Is the tendency for people to overestimate their ability to control events, for instance to feel that they control outcomes that they demonstrably have no influence over.
http://en.wikipedia.org/wiki/Illusion_of_control

If you want a lesson in how powerless you are as a CA, and I suppose as any human being in relation to another, then in meeting a new service user this is brought home to you in the most basic of fashions; you quickly realise that although the service user may be relying on you to provide many of the basics for their quality of life it is they who are in the position of control.

This may appear like a contradiction to what I stated above about all care work environments being CA led, and service users being all too open to manipulation, but it hides a subtle but vital distinction – CW's "capture" *a service, not service users*, it is through capturing a service that they come to control service users, not the other way around; indeed it is this very lack of control over service users themselves that leads CW's to "capture" a service; CW's invariably seek to fashion a service to their convenience directly because care, if administered properly, *is by its nature unpredictable and uncontrollable*. The fact of how little control CW's have in supporting a service user leads them to find ways in which they *can* control the environment, if not the actual service users themselves. It is immeasurably easier *to control a service* than it is *to work with a service user* – an inversion of how real and proper care should be administered – good CW's seek to work with service users and to ensure that the service promotes *their needs and requirements first and foremost and above all other considerations* (what is called in the jargon Person Centred Care or the unfortunately ironically abbreviated term PCP – but more on this later).

However poor or abusive care is where CW's – realising they cannot actually control the service user - seek to control *how the service delivers care* - manipulating the conditions around service users' care to effectively "routinize" it – in other words while a CA may not be able to control a service users needs and demands they can heavily influence, and ultimately control, how and when those needs are met and so ultimately exercise a large degree of control over any service users in this fashion.

All this I was blissfully unaware of over the long period that I was waiting for my CRB check to come through – nearly 2 months. No matter how much shadowing of care staff I had done none of it was much use on my first day "going solo" – that is the first time I was left alone to support Service User "A".

Of course I had read all about Service User "A" from his gargantuan file - their life's history in an episodic fashion. They contained various psychologists reports, a written history, charts of behaviour, briefings, notes of meetings, doctors and psychiatrist diagnoses, Local Authority documents, some police reports and endless reams from social workers and care managers.

Looking through these files was like trying to reconstruct an interpretable past from a haphazard collage; yet it also made sense in an appositely Foucaultian fashion - this wasn't history, it was archaeology. The life of this person that I was soon to be supporting did not follow a linear or consecutive fashion in the way a narrative would, instead you had instances, elements, happenings, actors, voices on a page, and the only consistent thing was that they all had reference to the same individual – Service User "A" – the name was the only thing stringing all these discrete and disparate parts together. Every piece of paper appeared to tell a story, but one in which the main protagonists' voice was deliberately absent – they were silent, dumb, reduced to an object that appeared to have no concrete essence.

Here was a hospital letter about Service User "A" at 20, here was a psychiatrist's report on them at 32, here was a social worker's assessment at 40, here was police report at 50; all of these things recorded something of note *about* Service User "A", something that had been worthy or notable enough to commit to paper, but nothing *of* them.

This was strange to me at the time (later experience has made me used to getting to know people through what is written about them) as it seemed like I was both prying and, at the same time, being mystified; prying because I was, through reading, party to details that most of us would consider not just confidential but deeply personal and perhaps never to be revealed to anyone; mystified because while I got to know a lot *about* Service User "A" I didn't know *who* they were, I was learning *what* they were but this came some distance short of knowing *them*.

I couldn't help, reading through all this material, wondering what files, reports and letters concerning me may be scattered over different files in different places concerning different incidents or events in my life, I wondered how I would be constructed through all these records, what someone would make of me if everything were laid out in front of them.

The thought also struck me that even the most balanced and uneventful life could be made, through this medium of partial and fragmented history, to look like a life of a madman. Learning about someone this way seemed to me to be also magnifying their disabilities – what of all the bits of life in-between these documents? What about all the days amounting to years that had been lived without being deemed to be noteworthy? What pleasures or sadness's that most of us consider as making up the most important parts of ourselves were missing here? I wondered did Service User "A" like the smell of newly cut grass like I did, did they like the autumn with its veiling mists and life-drained burnt gold leaves falling, did they dream of past days joys?

None of these things were recorded, yet how important they were in making a real picture of a person were inestimable, however, in this context, they were deemed redundant, beside the point, unworthy of recording.

Some of this was unavoidable, while there was nothing that told me what sort of person Service User "A" was they did inform me of the nature of their conditions and how they had been treated over time. As Service User "A" had, for most of their life, been institutionalised it did give me an awareness of how they had experienced life, and it is this, more than anything else in care that I found was the key to making myself a better carer.

All of our lives are to an extent synthesised through our relations with others, the very way that we experience the world is mainly directed through how we act and interact with other people, the people we are or may become is to a large (but disputed) part based on our relations.

The most obvious of these are those with our parents, they form and colour our world view which stays with us for the rest of our lives – even if it is only to react against. From our parents we get the sense of ourselves through our history both direct and indirect – they remember things from when we were young children that we do not and they bring with them their heritage – where we come from, where we fit into society. Positive or negative relations to our parents may shape all our other relations and so the way we experience existence. If though our parents are missing we look for other role models, substitutes or proxies; but in a society that is based on biological parentage the absence or lack of blood parents can be a disorientating experience. The lack of grounding, of sense of continuity or history is broken and often that very lack leads to difficult relations with others – how are we act or react if we are not sure where we actually came from?

If those who care for us from a young age happen to be professional carers this complicates things to a degree that very often vastly understated and therefore misunderstood – how would the most balanced of us (that concept in itself is subjective) feel if those to whom we were closest to when growing up were in fact *paid* to be caring for us – no altruism, no unconditional love, no sense of moral duty – only a pay acket at the end of it. This fact can be endlessly damaging, a fact I witnessed in both the people I came to care for and those I worked with too; trust here is almost absent, a high level of egoism replaces it and the overriding imperative is "what can I get out of this relationship" rather than "what can I put into it" – this isn't a moral judgement it is simply cause and effect – if the first experiences of relations are conditional so too will the relations of that individual be too.

Imagine all the conditions mentioned above then – a lack of a sense of history, lack of direct blood relations, deep trust issues - and then throw in severe mental illness and you have something approaching the issues that Service User "A" had to deal with daily.

Wider than this though you have the condition that *nearly all* service users face and which in turn affects their relationships with the CW's who support them. In supported living, almost by definition, a CA will encounter those whose attachment to family has been either suddenly broken or slowly severed over time. This fact is still, shockingly, overlooked in the holistic care of service users and betrays an attitude in care more akin to the early part of the last century than this one as it implies that because they suffer from some sort of mental or physical impairment that they do not have the same need for a grounding or history as the rest of us.

In fact their need is greater because in the majority of cases service users find themselves in a care setting because either

1) their families' cannot cope with their care needs,
   or
2) that their main familial carers have become too ill, old or have died;

Whichever is the case they find their real attachment to a definable origin broken; instead they perceive – rightly - themselves being supported by strangers who are not disposed to their care by duty but by contracts and payment. Care support for the adaptive mechanisms necessary for such a transition are often neglected leading to all manner of behavioural issues; that the root causes of these are not effectively tackled is one of the great lapses in care; too often the fact that a service user has ended up in care are glossed over for the sake of temporary peace but at the cost of long term welfare.

Perhaps worse than this is that it places individual CW's on the spot in regards to explaining to service users how and why they have ended up in care; as a result service users will often get several different answers to the question – "Why am I here?" As can be imagined - many different answers to the same fundamental question will only exacerbate trust issues. Few are the support structures that put in place a consistent answer to the question – "Why am I here?" yet it is perhaps the most fundamental question of all that any of us face.

In regards to care for the elderly in a residential setting the same issues arise. Just because someone has dementia does not mean that they have lost all capacity to ask just what they are doing in what is an often disorientating and disturbing new environment. In fact dementia care – because of the fact that is causes regression – the recrudescence of childhood memories and the loss of awareness of the sufferers own age – the need is greater.

In my years of working in care for the elderly "Why am I here?" is one of the most often repeated questions I have been asked, and the most difficult to answer, for how are you supposed to reply that they are in a care home because their family either cannot or will not support them. To the dementia sufferer it may come as news that they even qualify in age to be in a care home believing themselves to be vastly younger than they are; even more difficult though is when the sufferer is acutely aware of their age and their surroundings, here the fact that they feel abandoned by their families is immensely damaging. Frequently I have wanted some form of agreed response to this question that I knew would be reinforced by other CW's, instead I have often had to fall back on the my own stock response – that the sufferer has been ill and this is a place for them recuperate.

This lack of a clear message about origin and history may seem to be overplayed here but it is key to a CA building up a trusting and mutually beneficial relationship between the service user and the CA. The service user needs to be able to trust the CW's who supports them and the CW's themselves need to build a working relationship that puts the service user at peace and allows the best level of support to be provided. A caring environment cannot be sustained without trust and support, this takes time and is fragile, as I was about to find out, but first I had to rethink all I thought I knew about mental illness.

I like many others, even having suffered mental ill health for a lot of my adult life (and, if the truth be known, I believe from early childhood) I thought that I understood mental illness, that my empathetic qualities regarding my own experience of a mind coming apart at the seams would overcome all barriers between myself and anyone I was to care for. In fact I was as in the dark and as ignorant as anyone who may think that mental illness means "nutters". I, like many members of the general public, liked to think of mental illness in "silos" - that certain behaviours were linked to the underlying condition of the sufferer while others were not. This is a basic misconception and can be illustrated best in the following which is based on a real service user that I worked with.

The service user in question often refused to change dirty clothes wearing the same soiled items day after day. This behaviour was linked and "explained" by professionals and managers as an expression and symptom of a depressive episode in the bipolar disorder that the service user suffered from. The approach that I was advised to take was to encourage the service user to change but not to provoke any challenging episodes. However this service user also at times soiled themselves and smeared it all over all the reachable surfaces of their room. This action was classed as behavioural and therefore a manifestation of more prosaic issues– temper, attention seeking, anger, resentment, unhappiness, upset – that they felt that they could not communicate in more "conventional" terms. The approach for CW's to this was to try to get to the bottom of the underlying situation and to tackle it, (this, and of course spending several hours cleaning the service users room) it was also deemed best practice to communicate to the service user in clear, precise and unthreatening language that this "behaviour" was not acceptable from a sanitary and welfare point of view and that they should understand that the same result – the tackling of the root problem – could be achieved in a less anti-social manner.

Far be it from me to question the approach that had been devised by more senior CW's, the service user's care manager, service managers and assorted health professionals – psychiatrists, psychologists, the appropriate General Practitioner (the so called Multi-Disciplinary Team – MDT - approach) however it did expose a flaw that I believe exists in the treatment of *all* mental illness – how far can mental illness be said to affect *the whole of a sufferers life* rather than just portions of it.

Inadvertently, and sometimes arbitrarily, we look on some things as being connected to a mental condition and others as not; this is like isolating illness, as if it impacted at a particular point and nowhere else. This thinking is complicated in those with severe mental illnesses that affects their mental capacity as defined by under The Mental Capacity Act 2005 (MCA). As glossed by the mental health charity Mind the MCA deems that -

"Capacity is not a permanent status and so people should not be described as having or lacking capacity. Instead, when considering someone's mental capacity a health or social care professional should ask, 'Is this person, at this particular time, capable of making this particular decision?'"
http://www.mind.org.uk/mental_health_a-z/8059_mental_capacity_act#5

This means while in some areas someone may lack capacity, in others they are deemed to have it. The thinking behind this cannot be faulted as it prevents decisions being taken out of a mental illness sufferers hands wholesale and permanently and provides flexibility in apportioning autonomy by evidence rather than simply by condition. I would in no way challenge this but only would like to point out the flaw alluded to above that makes the understanding and supporting of a mentally ill individual so difficult. If we view some manifestations of actions in a sufferer as conditioned by their mental illness and others not we fail to grasp the nature of the effects of mental illness – particularly severe mental illness – in its entirety.

Yet to view mental illness as affecting only a portion of someone's critical thinking or actions is to misunderstand the holistic nature of mental illness. As a paper on the practice known as Empathic Witnessing makes clear –

"According to Edmund Pellegrino, disease is an "ontological assault" on the body."
http://www.inter-disciplinary.net/ptb/mso/hid/hid4/rosenberg%20paper.pdf

In plain language an illness is an attack on the very nature of being of an individual as well as an "assault" on the physical body – it has profound psychological effects beyond the illness itself.

Further –

"A disease can leave any body agitated and troubled, even if the person is of sound mind."

[Edmund Pellegrino was an expert both in clinical bioethics, and in the field of medicine and the humanities, specifically, the teaching of humanities in medical school, which he helped pioneer. http://en.wikipedia.org/wiki/Edmund_Pellegrino]

Although the paper concentrates on physical illness it also is easily applicable to mental illness – it is something that colours and alters the whole mental processes whether directly affected by a mental condition or not. This is not the same - as I have pointed out in other areas – as defining someone by their diagnosis, in fact it is the opposite - it is finding a person behind and yet involved in their illness.

A person does not disappear into their illness but no understanding can be reached without taking into account the immensity of it.

What the hell, you may ask, does this have to do with supporting someone with a mental condition or illness like dementia? The answer is quite a lot. No CW can do their job with any degree of compassion and competence unless they appreciate that *everything* a sufferer does is *to a greater or lesser extent* conditioned by their illness. To put this in more practical terms let me go back to the example made above. Both the refusal to change clothes *and* the liberal smearing of faeces are manifestations of mental illness even if the latter was deemed behavioural. What is required in either case though is that CW's must try to break through the subjectivity of each person they support in order to better understand and therefore care for them.

The smearing of faeces in the above case is a manifestation conditioned by the service user's mental illness; CW's though were encouraged - by the different classification of this action to treat this manifestation differently – they were to see the refusal to change clothes as something that the service user "could not help" and so to be gently managed while the smearing of faeces was to be seen as something the service user had a degree of control over, that is, something deliberate. It inevitably follows from this that CW's reactions to both these are different – the latter will be treated much less sympathetically than the former. This creates a breach in the relationship between CW and service user.

This also begs the question that plagues CW's, and more specifically, the management of them – does it really matter two farts in a windsock what CW's think.

This is the root of many a problem. Many members of the public and most managers will take the view that it doesn't matter a fig what CW's think as long as they do the job. This is an idiot's response and should be treated as such. In fact it is critical what CW's think as it colours the whole of their quality of work. A care strategy that leaves the door open to thinking that a service user has full cognisant control over some aberrant actions leads automatically to a lack of empathy; a lack of empathy leads to a drop in the quality of care; and a drop in the quality of care is only a short step away from potential neglect and abuse.

The main points to draw from this is that a lack of empathy is the death of good care, and to emphasise this point I'll put in capitals – ONCE EMPATHY IS LOST PANDORAS BOX IS OPENED TO ALL MANNER OF DEFICIENCIES AND DISASTERS IN WAITING. The second and perhaps primary point is that mental illness cannot be compartmentalised, cannot be isolated; mental illness affects everything a sufferer does and every action undertaken by them should be viewed in the light of this underlying condition.

I would be lying if I said that I had never fallen, in my time in care, into this trap of lacking empathy and compartmentalising mental illness. There have been numerous times when I have been told that a service user's particular action was "behavioural" rather than evidential of mental illness, it was then impossible to view that behaviour with the same compassion as I did with other actions. Indeed I found with such actions a shortness of patience and consideration that at times surprised even me, and an inability to approach the results of these actions with any form of equanimity.

Time and again on leaving work and looking back on my efforts I lamented that I had not acted differently and made up my mind to approach such actions with a better understanding on my next shift, only to find myself back at work falling into the same pattern over again. For the truth was that once the seed had been planted that a service user could have some degree of control over their actions, and further, that they chose certain actions in order to gain attention or to otherwise express emotions or feelings, it was almost impossible to behave differently as a CA.

This may be a shocking admission to some but should really come as little surprise, CW's are only human and as such are susceptible to cues, signals and influences as any other human, therefore it should not be a shock when we act like "other people". If someone you knew had a medical condition that made them abnormally flatulent you may be able to deal with this, however if they suddenly urinated all over you what would your reaction be? This is not as ridiculous as it sounds, and is of a similar order to that which many CW's find themselves faced.

Building up trust with "Service User A" encompassed all of the above and was key to formulating a lot of the thoughts and conclusions I have drawn in the preceding, it also emphasised the enormous responsibility that was so easily cast upon my very inexperienced shoulders in that "Service User A" would be as dependent on me for the quality of his day as I would on him to dictate how I was able to best support them. The learning curve was steep but had to start somewhere.

CHAPTER VII

Disturbing

Perhaps it was my age, or maybe it was from the prolonged period out of work, or then again it could have been the enforced temporisation incurred by the prolonged wait for my CRB to come through, but my first day on my own with "Service User A" was an unusual one. On the one hand I was almost giddy with the prospect of finally becoming a "full" member of the care team, to "have my wings" so to speak and to be placed on an equal footing with them, I felt like, at last, I was making progress in my new career.

Then on the other hand I was almost sick with the anticipation of the moment when I would be left alone with "Service User A" akin to the deep, partly physical, part psychological nausea that reminded me of the first day back at primary school after holidays. I was struck with the thought that I would be on my own, thrown back on my own resources and left to manage as best I could. Of course in such a febrile state my mind was doing flip flops as I made my way into my first lone shift - part of me saw myself as being dynamic, alert, inspired and able to unlock hitherto unseen potentials in "Service User A" that my brilliant support work would enable. The other part of me saw me run out the house with all the windows put out, the furniture smashed and general apocalypse unfolding because I'd said or done the wrong thing.

Both of these scenarios, in retrospect, were each grandiose in their own way. The fact, as noted above, was that as a CA my influence would be pretty limited; whatever I did or did not do could improve or be deleterious to Service User "A's" day, but it could not, by and large, influence whole outcomes or the thread of their life. The truth I was to rapidly discover was that I was there in the main to simply to enable, at times possibly to benignly influence, and maybe, just maybe, at times to extend the boundaries in the quality of "Service User A's" day, but for the most part I only had to *be there*, to watch, to listen but mainly just to be as unobtrusive as possible.

Yet the overriding feeling that I had – viewing potential scenarios both positive and negative - was of the weight of responsibility, the feeling that I was, in however much a limited fashion, in charge of the care of another person.

In care perhaps this feeling only happens once to CW's - right at the start of their careers - and is quickly lost in the hustle-bustle of absentminded routines, details, paperwork and the ephemera of care; yet maybe this is also a limited state of grace. Care, ultimately, is the looking after of the welfare of another human being who cannot function satisfactorily without support. Little in life, I thought then, *and still think now*, can be of more significance.

Much is made of parenthood as the guardianship and responsibility of a vulnerable other, yet care is of a similar order, with one major difference though – it is paid job. This obviously critically alters the dynamic at work; there is no familial bond, no link in the chain of existence, no obligation of blood. In fact - and stripped of all the societal and PC niceties - care work is a financial transaction – a service provider purchases the time and expertise of a carer who in return is obligated to offer at least a baseline of support to a service user.

This antiseptic view is in reality how most CW's come to view their "job". I was no different later in my career in care - the lustre of that almost dread that I felt on that first day having long since fled – and left me looking at my caring role as a duty owed to my employer for the wage that I was paid, not a duty toward individuals to improve their life's' circumstances. This does not mean that I stopped caring, but I forgot how intimately and ultimately responsible I was for those I was there to support. Occasionally I got glimpses of how responsible I felt (usually when something went wrong and I thought I had failed in my "duty of care") and how much of service users life quality depended on my really caring, even at one professional remove, but the visceral feeling had gone, all I was left with was a shadow.

On that first day though I felt all the weight, so much so that 20 minutes into my shift, without any incidents or untoward happenings, I was as near to a full-on panic attack as I have ever been. Frankly I was terrified. Plenty of deep breathing later I managed to pull myself together. On reflection I wondered how other CW's on their first day in their first job in care coped, if I, a man of nearly 40 years of age and life experience nearly crumbled. After all I had not been thrown in at the deep end; I had had plenty of time of observe and study other CW's working with "Service User A" and couldn't have been more prepared. What of those who's CRB's had come through in a matter of days?

I was sometimes thrilled, more often stunned (in a good and bad sense) and occasionally repelled at how lightly young and/or inexperienced CW's took on such responsibility; any which way they seemed to cope with the experience with much less trepidation than I did.

Much of the insouciance I observed in younger CAs could be put down to the fearlessness of youth; that things which faze those of us of more advanced years barely seem to trouble younger minds. This is one of the joys of being an older care worker, to look at how adaptable and capable many young CW's are, how readily they absorb experience and use it not as something to be chastened by but as something along a learning curve, how they are able, without pause, to walk with confidence into many tricky situations that with age brings only agonised hesitation.

Yet there is a corollary to this, at times such fearlessness also indicated a lack of thought, a lack of circumspection and a failure to grasp the gravity of certain situations. Sometimes age brings well warranted fears, trepidations, circumspecton, things that take into account the consequences of any actions and – even more importantly - the sometimes even greater consequences of inaction.

Having no fear sometimes means having no thought either.

For all the rewards of seeing how younger carers were able to quickly assimilate information and flourish into fantastic carers I have also seen many that seem to fail to understand the gravity and importance of certain tasks or jobs. As I gained in experience I found it often frustrating that some of younger colleagues thought that what care meant was simply making sure the service users they were supporting didn't die and that was all. Sometimes it didn't even reach that level, they thought that by simply turning up was the whole job done. Although such carers were mercifully rare they were not so rare as to be of no significance. In many cases I have seen a carer do as little as possible even when this means the people they are supporting are left in discomfort and distress. Often this was down to young carers simply not grasping that they were involved with people who depended on them, but just as often it was down to just *not caring*. Good young carers' are a great asset, bad ones a limitless liability; but in-between is the area where some balance has to be struck between their fearlessness and adaptability and their folly and sometimes their reckless disregard.

Back to my first day though.

Nothing dramatic transpired that could said to be noteworthy but much happened in the indefinable space that exists between a CA and a service user that they are supporting. In some ways it was like a circling of two wary species encountering each other alone for the first time, a shadow boxing match between two protagonists who never quite meet, a silent movie with no title explanations. "Service User A" tested me a couple of times, I had not been quick enough getting his meal from suggestion to table, couldn't load his favourite movie on the DVD player, asked him too much if he was ok when all he wanted me to do was shut up and be there, but mainly he just tried to ignore me as if I was another irritation in his life (which I – in that awestruck first day – probably was). This was when I truly realised how much more in control of their day "Service User A" was than I; how they could dictate to me how comfortably or otherwise our day was to be spent together.

While some may feel I am crediting too much nous to "Service User A" I distinctly had the impression they were challenging and experimenting with me so as to find out what kind of CA I was to be in relation to them - was I going to "push back", assert myself, try to impose my will? What latitude was I going to offer them? How far they could stretch me? Each action calibrated to see if I would stress or break. Although I wobbled I didn't do any of the above, what I tried to do was to be as "obedient" as possible while setting boundaries – I apologised when I could not meet their immediate demands but also explained why those demands could not be met as immediately as they wished.

What in fact was occurring was the slow building of something and that slow built something was the beginnings of a relationship; I was being tutored by "Service User A" in the way that he wanted to be supported.

This was, and remains, one of the most rewarding factors I was to encounter in care - how two complete strangers, meeting on totally artificial terrain and in a burgeoning relationship that involves complex webs of powers and controls, could be somehow be conducted on a totally human level; for over time not only did I get to know "Service User A" I also came like him in a way that I would have thought unimaginable at the outset. I came to see that "Service User A" was funny, quick, unimaginably intelligent considering his outward appearance and presentation, complex in the ways that we are all complex and most frequently heartbreakingly loving.

Nothing could have illustrated this final fact more than when I managed - after two weeks trying – to be allowed to enter the sacrosanct space of his bedroom and change his bedding. As I pulled back the sheets I noticed something small, ragged and hopelessly discoloured; at first I thought it to be an old handkerchief, instead, on closer inspection, I found it was actually a stuffed bodied doll about 15 - 20cm in length lying next to the dent in the pillow where Service User "A" had obviously rested his head. There was nothing sexualised in the doll, nothing fetishistic – it was simply, with its frayed but still intact body and grey dress (darned by who knows who in some places) a projection of a love that had never had had any proper object. This brought me right up against what Pelligrino called "wounded humanity" - the fact of the human crouched and cowering inside an all-enveloping illness. Alone in his bedroom I unexpectedly felt tears start to run down my face; cringing with inner shame I wiped them away and placed the doll sat up and carefully arranged on the one chair in the bedroom while I made the bed and then carefully tucked the doll back under the covers lying on the pillow much as I had found it. I felt guilty for having disturbed it, "Service User A" had evidently forgotten it was still in bed so hadn't hidden it in anticipation of my entering. Somehow I felt more than a little guilty too, as if I had violated something that had been kept back from all the files and all the advice and all the experience that had been passed on to as to how to best support "Service User A", as if I had blunderingly invaded a small space that was theirs and theirs alone.

Try as I might I could not get that doll image out my mind (indeed it is as vivid now as I write this as it was when I first encountered it) and what it represented. Where had the love that is normally projected at objects of affection gone? "Service User A" had never had a family background; he had been abandoned to the care system from an early age and had probably spent the majority of his existence – could it really be called a life? - kicking round the backwaters of it for years on end – his copious notes and history were strangely silent about the period between him being committed to the care system and his much later years in a secure hospital when he was already well past middle age.

Probably cared for by a succession of transient health professionals they had never had anything approaching stability never mind comfort or security. Yet where did all that capacity for love go? Where did it find its own home? This much loved and, by the looks of its tattered appearance, far travelled and long possessed doll was the only evidence of a close and loving relationship. Nothing had been said about love in all the huge tomes I had read, nothing about attachment or affection.

This brought home to me a fact that still we as a society remain squeamish about the mentally challenged, or severely affected, being sexualised beings. We like our mentally "disabled" to be asexual, devoid of drives that many of us consider normal, indeed vital; we want them to be "brave" or "loveable" but we don't want them to be sexually charged and loving in a carnal way – when they are we choose to either look away, question it or – and this is if we are all honest – be rather ill at the thought.

We like to think we have travelled far from shutting off mental disability from society in institutions or closed communities and yet we still deny something so central to humanity – physical love - is this just not another kind of ghettoization? A more important issue than even this though is if we deny sexuality we also deny other capacities to love in other ways; we can have no conception of the capacity, or more importantly the need, of someone to love in a caring sense if we cannot also conceive them as sexual beings also, we as a society often persist in this denial of sexuality.

Within the care environment this relationship becomes ever more confused as the notion of protecting a service user comes into the equation. CW's are first instructed to be risk averse, to not allow a service user to enter, or be drawn into, an unacceptably risky situation; any risky situation that CW's cannot avoid exposing service users to has to be carefully and minutely managed by a risk assessments and then cleared by management and more senior care officials so that the risk as opposed to the rewards from the activity are acceptable. But how do you risk assess human relationships or the expression of sexuality? As any parent knows, to allow their children to love means also to allow them to be hurt, yet is it possible to quantify hurt if an individual has trouble communicating their emotional interior? Often by this denial of love and care within an individual its corollary – sexuality - is forced out in ways that society deem "inappropriate" or even transgressive. This poses sometimes impossible demands on CW's supporting service users both in the community and in residential care, two instances will make this clear.

Another CA I spoke to in researching this book supported a client who was considered a potential danger to minor's. This was based on an incident in the service users' past where they had exposed themselves to some children and attempted to masturbate. This service user too had been in the care system for some time and institutionalised, they had never had a "normal" relationship although they were intelligent and able and therefore were aware of this lack in their lives. Although no police action had been taken over the incident it was part of the service users care plan that their behaviour should be monitored as to all sexualised behaviours - they were to have exclusively male carers or only female carers above a "certain age" judged not to be "stimulating". All images on TV or in papers or magazines were to be vetted thoroughly for any sexualised content and removed if they were deemed potentially "dangerous"; and in any interface with the public they were be kept under close watch by two CW's (within the service user's own house one-to-one care was deemed sufficient).

The whole situation was impossible to unravel as to what extent the CW's were being asked to protect the public from the service user, to what extent they were protecting the service user from their own transgressive conduct and to what extent they were actually facilitating a continuance of a truncated and therefore dysfunctional sexuality that may itself have perpetuated that same transgressive behaviour – this is a point we shall come back in more depth below.

There is though, first, a more germane point here in regards to how this service user was supported that returns us to the main point made earlier about inexperienced, inattentive or uninterested novice support workers. It should be obvious from the above that a key task of those CW's supporting this service user was public protection – yet it is stretching the bounds of the baseline of experience and expertise of support workers that they should be placed in this role. Support workers are meant to be just that, CW's who provide support to enable a service user to function with the utmost possible independence - they are not specifically trained for a public protection role.

Fortunately (and it was down to fortune, not planning) this service user had a good, mature and experienced regular care team, however it was equally as possible that they could have been supported by younger, less experienced and less diligent carers. In this situation, as noted above, the lack of awareness of the potential gravity of any lapses in their care - and the consequences of it - could have had catastrophic results for others and for the service user themselves (a point I shall return to in a moment). More troubling though in this instance was that this service users regular carers were sometimes moved service during periods of staffing stress, and this meant that the care of this service user, and therefore the protection of vulnerable others, was often in effect committed to a lottery, as at times bank shift carers (qualified carers not on full time contracts but able to be called upon as and when they were needed) were allotted to care for this individual, many of whom had no prior knowledge of the sexual issues and dangers at stake.

This not only put vulnerable members of the public at risk it also put the liberty and care of the service user in jeopardy, for the care team were also protecting this service user from themselves. As the service user had no conception that their attitude and sexual leanings were transgressive in the most aberrant fashion it was impossible to communicate that how they were expressing their sexuality was not only inappropriate but criminal – not least because even discussion of sexual issues were made taboo in the care plan – so the care team had a duty of care to the service user to protect them from their own actions.

Although protection of the public and protection of the service user coincided here once more CW's were being pushed right to the limit of what could reasonably be expected of such lightly trained care professionals – that they support the service user, protect the public and protect the service user from potential dangers in their own actions.

Then there was the issue of the removal of all potentially stimulating images from the service user's view. Although there were sound reasons for this – the fantasisation of actions fuelling the desire to fulfil them so accelerating and heightening drive toward the experiential – it also maintained the de-sexualised "bubble" that had possibly contributed to the "deviancy" in the first place. The removal of all sexualised images meant that even pictures and representations of normative sexuality were unavailable. The difficulty here is obvious – here was an individual with sexual drives which were channelled in an inappropriate way but which had no way to being expressed in more appropriate ways, a (rightful) denial of one expression of sexuality lead to the denial of them all. This placed this service user in a situation that would be ridiculous if it wasn't tragic and to some extent cruel; the only behaviour that could be tolerated was one that denied their own sexuality.

The CA I spoke to saw the absurdity of their position – enforcing a regime that sought to cut off at the source deviant behaviour but by such actions possibly exacerbating it – they also had a theory of why this service user had been sexualised in this way. As no one really gives a shit about what CW's think this carer didn't express their views professionally but only off the record to me.

Their theory was that having had all potential loving relationships denied by circumstance, and the care system itself, their normal need for expressing love and seeing it reflected back had become infantilised and confused with sexualised love - it was love in search of an object that had collided with sexualisation – the result being deviancy.

This only reinforces the point that the denial of sexuality is infinitely damaging and ultimately dangerous – the fact that we take that as read for "sane" individuals but do not apply it to those considered to have learning disabilities only emphasises the point made above – that as far as we have come in bringing disability out of the ghetto, our minds are still somehow lodged there when it comes to feelings and expressions of love.

Be that as I may this analysis in regards to the service user in question is moot – having come from "only" a CA who had worked intimately and daily with this service user for quite some time - the actual management and care for the welfare of this service user was - and remains at the point of writing - left wholly in the hands of more highly qualified health professionals, clinicians, psychiatrists and psychologists together with senior care managers inside and outside the service provider.

This would be all well and fine were it not for the fact that while the very views of those perhaps most acquainted with this service user – the CW's who supported them - were not solicited, they were in fact wholly and wilfully ignored – it was they, and not the more highly qualified health professionals, clinicians, psychiatrists and psychologists and senior care managers inside and outside the service provider, who were left to pick up the pieces of this sexualisation; they were asked to do a job right at the limit of their capability in order that care objectives could be met without their valuable experience being taken into account.

This is the ultimate tragedy the CW's position – that they are asked to implement a regime in which they have no input and therefore no stake – yet they carry the onus of any consequences that result from this care regime. While it is only right and what is best about care as it is currently configured that more of it is provided in the community, care in the community can only be safe – for others and for those being cared for – if the management of their condition and the manifestations of their illness are adequately catered for; simply to push the onus on CW's with little or no training and support is unsatisfactory, it is like throwing someone out of an airplane and asking them to knit a parachute on the way down.

Another incidence of how sexuality can become a point of major difficulty for CW's occurred in a residential setting. Another CA told me about the situation they found themselves in. In this service several young adult service users with more or less severe learning disabilities were housed in one large building. One of the female service users was on a strict calorie controlled diet due to other health issues beside their learning disabilities. To this end their blood sugar and weight was monitored daily and weekly respectively. Over time, and despite careful monitoring, this female service user was gaining weight and at times their blood sugar was elevated beyond what it should have been for the diet they were on. Their bedroom was searched for any secret stash and their food intake monitored ever more closely, all to no effect. One night the fire alarm went off and all the service users had to be evacuated. The alarm was a false one but another service user – a male one who had a high degree of capacity and ability - was discovered in the female service user's room which was littered with sweet wrappers. It transpired that the male service user was bribing the female service user with sweets in order to gain sexual favours. Once this was discovered the question then arose as to what to do. As both of the service users had mental capacity as to their own choices – that is the ability to make all but a few key decisions about their welfare - there was little that could be done. Even though the sexual consent had been effective procured by means of sweets this was not deemed to alter the female service user's capacity to decide to engage in sexual intercourse – they were not intoxicated, they were not rendered helpless and, by the admission of the female service user herself, she had choice but chose to engage in intercourse. The only issue that any action could be based on then was the pure health consideration about the sweets. The male service user was admonished for breaking this prohibition but could not be for anything else. In the end it was decided that the female service user should be moved to another residential unit on grounds of health and welfare and she was offered contraception.

The first thing that springs to mind here is that the male service user was abusing the female service user. In the broadest definition maybe yes, he certainly was engaged in a morally equivocal actions and the element of "bribing" could have been the crucial factor in procuring consent from the female service user, yet to call it criminal – and to prove it - was impossible unless the female service user was judged to lack capacity, which she was not. To make a rough analogy would it be considered a police matter if a male alcoholic procured sex from female alcoholic by means of the promise of a drink (note here I am not suggesting prior intoxication) and the female stated that they freely consented on these grounds? In this scenario the man may be acting horrendously, exploitatively and in a morally degenerate fashion but it would be impossible to prove any crime.

The same fundamental issues existed here, the only difference being that the CW's that supported these service users were supposed to have prevented this exploitative situation developing. In fact the axe only really fell on the CW's in this instance as three of them were fired the very next day after the discovery – the grounds – that they failed in their duty of care toward the female service user. The rapidity and severity of this action was due to the fact that as the Local Authority Safeguarding Adults Board had to be brought in the service provider needed to show heads on pikes to satisfy them that remedial action had been taken. The heads on pikes being all CW's.

This neatly illustrates the dilemma that many CW's find themselves in – that they are responsible for the greater welfare of those they support even when the actions and choices of those they support are undertaken within their capacities. Quite what the CW's were supposed to have done here is difficult to construe; yes they may have been more diligent; yes they perhaps could have spotted manifestations of questionable behaviours sooner – although quite what these would have been I cannot say – and the night care team perhaps should have likewise been alive to potentially compromising situations; but this does not alter the fact that if they had been more alert, if they had perceived and stamped on the male service users' actions more quickly, then they were in fact denying choice and freedom in sexual matters to the female resident.

Some will argue that this is a bogus argument and that this issue transcends choice and comes under the remit of exploitation and abuse. There is legitimacy in this but raises the further question of how far should choice be extended and where it should be curtailed in – as in hackneyed language of care – the service users "best interests" (best interests is something that we shall meet again in dementia care).

What are those best interests though? Are they not a projection of our own perceptions of what is and is not in the best interests of the service user? And, to return to the parent/child relationship, is it the responsibility of parents to police all of the potential sexual liaisons of their children and for them to judge what is the "right" or "wrong" choice. Who among us – male or female – have not had experience of sexual relations that emanated from warped or bad motivations? and is it not the bad choices that we make that are the ones that lead us on to different and better choices? If we were to be denied all "bad" choices in the sexual field, as adjudged by our parents or other adults, then is it not an obvious consequence that we will be bound to make ever more disastrous choices having been denied the learning experience that normally would arise?

All these are difficult questions when we introduce the element of mental disability and consequent capacity; yet to avoid them is a failure on society's part. It is not good enough that we push responsibility onto CW's and expect them to formulate answers or to be held responsible when those higher up the responsibility chain make decisions that place CW's in impossible situations. The challenge, I would argue, that now faces adult social care is to view those we deem "vulnerable" or "challenged" as "whole people" not as objects or subjects when it comes to the sexual field of play, and that we weigh free decisions more carefully in respect to their autonomy. There are no easy answers which makes it all more pressing that we as society decide to "piss or get off the pot" – either we take those with learning disabilities as "whole people" or we marginalise them.

The consequence of not treating those with learning disabilities as whole people was made manifest in my time supporting "Service User A" and was one of the most depressing and emotionally bruising experiences I had in the care system. I could sense the feeling of wanting to love in "Service User A" but having no direction to take it in other than that ragged old dolly he slept beside each night.

I could also see with the evidence of my own eyes how devoid they felt their life was as it was displayed when "Service User B's" parents came to visit, as they did at least weekly. This was the only time that

"Service User A" showed any challenging behaviours while I was on shift.

When Service User "B's" parents called "Service User A" made themselves scarce, closeting themselves in their bedroom, playing their music loud and occasionally throwing things out the window. "Service User A's" mood took the rest of the day to improve or for him to interact with the care team. I could only wonder at the level of hurt they must have been feeling – to see someone else loved and cared for by people close to them and yet to have all that denied them for reasons they could not conceive of. I found this aspect difficult, "Service User A" was now getting on in years and I wondered what had been quality of the life they had already mostly lived. Try as I might to stop taking an overview I could not help feeling they had been short-changed by arbitrary existence and circumstance. Yet in some ways it pushed me into being a better CA with them, whatever life they may have had up to this point I thought that it was now my responsibility, a day at a time, a shift at a time, to make what life they now lived as good and as rich as possible. The issue was I couldn't always be allowed to do this by the barriers raised by the care regime I worked in. It was this frustration and inhibition that eventually drove me out of supported living, but for now I did as best as I could.

If the lack of love and close relations were a wounding part of my care experience then with "Service User B" I was to discover the major drawback in having this support.

CHAPTER VIII

Relative Values

Quite the trickiest part of care for any CA is accommodating the desires and preferences of the relatives of those you care for or support. Often this relationship is characterised by barely concealed antagonism and competition that makes it feel like you are striving against, rather than working with, relatives; the reverse probably also applies in spades.

The most frustrating part of this relationship is actually untangling what the relatives *think* is best for the care for loved ones based on raw emotional attachment, and what is *actually best on a day to day basis* from a holistic care and medical point of view for the people you support.

It is an unavoidable fact that you - as a carer - will see more of those you care for *as they are now* rather than what they might have been in the past. Relatives will often base their assumptions and desires as to what is best for those you care for on information and experience that is years old and takes no account of either development and growth or recession and decline.

Frequently you, as a carer, find yourself partaking in the most absurd of practices - not for the benefit of the service user - but for the benefit of their relatives. This can often lead to damage and harm to a service user in respect of the quality of care you offer that may come to represent an un-healable breach in the care relationship. There is often nothing more likely to send CW's into either paroxysms of rage or a glazed over catatonic state than relatives appearing with a list of massed complaints, ridiculous requests and absurd demands that they expect to be fulfilled by this time last week. At times it feels like you are providing care *for the relatives,* rather than the service user.

Further families will often take horrendous risks and damaging actions with their relatives that if any CA were to undertake would result in severe censure or worse from those very same families.

Thrown into this toxic mix is the low esteem that CW's as a whole are held by society at large. Families inevitably have preconceptions about CW's based on the numerous horror stories that are often daily splashed across local or national media. Almost all families now regard care and carers from a default position that they are bound to fail in the quality of the care they are meant to provide and are therefore sensitised to the least mistake or misstep and expect to find evidence of poor or absent care everywhere.

Therefore the care relationship with relatives of a service user are usually hobbled and defective right from the very start.

The negative expectations, the immediate suspicion and, what is often an overlooked aspect, the differences in class, race or background between relatives and carers will be looked at in more detail later; firstly though I should like to address not simply what families bring to the relationship – over which CW's have no control - but the responsibly of CW's and clinicians in how they manage the care relationship with relatives - which is key to providing consistent reactive and cooperative care to service users.

Without a doubt the burden of understanding of what is at the root of a family's desires and wishes in respect of the care for their loved ones lies with the carers and clinicians within a service; for simply to ignore or summarily dismiss issues raised by them is self defeating of good care objectives as well as unhelpful, ill-conceived, chronically disrespectful, and not to mention frightfully ill-mannered. While not agreeing with *all demands* a family may make, a carer, or at least a good one, should try to understand the causes of these interventions.

At heart is one ignored principal that in reality should be placed front and centre of any interface between care staff and relatives is that service providers not only care for the individual service user, *but also that service users family.*

This point is worth making again and emphasising – HEALTHCARE PROFESSIONALS DO NOT EXCLUSIVELY CARE FOR THE SERVICER USER BUT THEIR FAMILIES TOO.

By this I mean that they should see the family as an extension of the service user; they too will need support, understanding, interpretation and compassion, they too need to treated with respect, decency, and have the right to make informed decisions in regards to their safety and wellbeing; to fail to see that families need all these mechanisms of care is to create a fissure in the care relationship with the service user via their relatives.

A happy well supported and involved family will almost always result in a happier service user, the opposite – a family that is excluded, misunderstood, ill supported and unhappy - will almost inevitably produce a less than happy service user. It is therefore in the interests of CW's and other care staff to tend to this relationship just as carefully as that which is nurtured with the service user themselves.

No should underestimate quite how difficult a decision it is to place a loved one under the care of detached care professionals no matter how good they may be. Any number of feelings and emotions are crushed underfoot when a family decides that they can no longer offer the care that their loved one needs and so must turn it over to others. Failure, inadequacy, guilt, shame, anger, distress – these are only some of the feelings that families may labour under when they finally admit they can no longer cope with the care responsibilities that they have until now dealt with; of these emotions most of them will, at some time or another, come to be aimed at the very carers they have chosen to look after their relative. This is understandable, no one can lumber under the cosh of such feelings without being driven to distraction or worse, so it far easier to project these outward rather than for them to constantly aimed as an inner rebuke. Much of the opprobrium that relatives aim at carers or the service provider comes from these sublimated emotions.

Simply to dismiss relatives as suffering from any of the aforementioned though is to simplify to the point of irrelevance the infinitely more complex dynamics at work in and through families.

One of the most important and vital points for a carer to understand is that by the time a family finally comes to the decision to place a loved one in full time care they are more than likely - individually and collectively - suffering from mental, physical and emotional exhaustion.

In my time in care I was almost always astounded - not that families had placed one of their relatives in care - but that it had taken them so long for them to do so. Frequently I saw new residents come into a service with so many issues and challenges that I was literally stunned that a family, any family, could have coped with them for any length of time. As a professional carer I was often run ragged by certain service users and left at virtually the end of my resources, so I could only imagine what it must have been like for the family. Under such long term stress the main carer or carers in the family sphere will often themselves be on the point of a breakdown or other such mental illness, therefore it is not surprising that they may come across to carers as being almost as irrational as the relatives they are placing in care.

A further critical point for carers to recognise is that within a family unit the care that will have been provided for a relative has often been one-to-one; the main carer will have had their relative alone to support. However once placing them into a care situation that care ratio (unless you happen to be absurdly rich or have platinum plated health insurance) will be impossible to maintain; in fact the care ratio will often go from one-to-one in a family to 1-7 or more; that is one carer for seven other service users besides this new addition.

It is almost beyond the obvious then to state that the same level of care cannot be provided in a residential care setting as it was in the home. Yet it is vital for CW's to understand that for families other residents or service users simply do not, to all intents and purposes, exist; only *their relative* is important.

This obviously creates a conflict as CW's are frequently directed by a family toward one particular service user at the expense of other service users and other care tasks. The difficulty of meeting several relatives' demands can create disruption to a service that costs other service users sometimes even the minimum of care. This situation produces a vicious circle – the demands of one service user's family will absorb precious care time from other service users who's relatives in turn will understandably ask why all their relatives needs and demands are not being met – and so on right through all the service users in a service. Having been in such situations the pressure and difficulty this poses is insurmountable and at times leaves CW's in a lose-lose position. It is a balancing act whereby CW's will have to consider both care priorities in regards to the service as a whole and assessing the immediate importance of requests made by a particular family, this means that CW's must be diplomats as well as carers – not all requests and demands can be met immediately so CW's will have to deal delicately with concerns, being both serious in addressing them but also in judging whether the request needs to be met immediately in respect of the care needs of all the service users under their care or whether it may be best dealt with later.

This calls for yet another facet that CW's find themselves having to cultivate – the engaged but professional carer who understands both the needs of the families and the needs of the all the service users under their care and prioritising care needs whilst still putting families minds at rest that their concerns will be met - but at an appropriate time and in an appropriate fashion.

This process for families – the change in understanding needed to realise that care can no longer be delivered on a one-to-one basis - is complicated by the attendant loss of control that the service users' main carer(s) will feel once they have placed their loved one in full time residential care. Understandably they will want to, and will try to, keep rigorous control of their care, even from a distance. Many of the minute demands placed on me by relatives have nearly all emanated from this admixture of the sharply increased care ratios and the loss of control. Often I have received instructions down the phone from families who seek to dictate care needs and requirements for their relative even while they have no idea of the present state and presentation of their relative. Many more have been the queries that I have fielded, often late into the night and into the early hours, as to the condition and wellbeing of their relatives.

In the most severe cases this attempt to reassert or maintain control has led to conflict over care provision. I have lost count of the occasions where long, difficult and painstaking headway has been made with a challenging service user only to see it all overturned in a moment by well-meaning but misguided relatives. This can be from the trifling – just having got an agitated service user settled only to see them needlessly unsettled by fussing relatives – to the serious – complaints and allegations against CW's that, despite being groundless, have to be investigated. I have seen several CW's who have done an almost faultless job suspended and placed under the enormous stress and strain of an investigation, sometimes involving the police, on matters that prove to be totally without foundation.

The transition from care at home by a family into residential care by health professionals is the most key phase of the care process - both in regards to the service user, but, most especially, in regards to the family - the respect, sensitivity and – as much as possible – the seamlessness of this transition can ultimate come to form the basis of the relationship between a care team and a service users family; handled well it can create the foundation of a trusting and cooperative relationship, handled badly and consistent, integrated and collective care can be dysfunctional from the outset.

Again it needs to be stated that, although it places additional demands and stresses on CW's, *the onus is on them* and a united care team – including clinicians and service managers – to see that this transition goes as smoothly as possible.

How frustrating it is then that in my experience whereas CW's have provided all the help and support necessary to make this transition work, all their good offices have been undone by tin eared clinicians and service managers. Countless are the times where myself and my colleagues have spent time with anxious relatives assuring them that we will do our level best in meeting both the new service users' requirements and the additional demands placed on us by those relatives, only to see all this fall apart by decisions and actions of managers and clinicians. Nearly always clinicians and managers have shown both disregard and frequently outright disrespect to relatives and effectively have sought to cut them totally out of the care process.

This dysfunction nearly always emanates from the top, the problem for CW's is that it is they who have to pick up the pieces of this very dysfunction.

This is illustrated in the fact that the main issue that CW's will need to comprehend and understand when in contact with families is that they will know much more about the service user than any CA, no matter how good the CA, no matter how much information the CA may have, no matter how long the CA has worked with a service user. This fact is often missed by carers because of the obvious paradox with what I have said at the head of this chapter – that as a carer you will know much more about a service user *as the personality they are now* rather *than the personality they might have been in the past.*

The contradiction though is not impossible to resolve; yes a CA will know more about a service user now, but this does not make up for all the time that has already past. Families are tied by shared blood and common history to each other and as such their bond is all the closer and all more intimate. A family may well have long memories of the relative before they became unwell, or else they may have raised that individual and loved them through their challenges and disturbances and therefore understand them on a deep emotional level far more subterranean than the much more superficial care relationship. Unless CW's understand this, unless they admit that while they may know an individual on a quotidian level they cannot ever fathom the deep and strong bonds that exists within a family unit they will be bound to come into conflict with families of the service user.

The shame and horrendous failing in care though is that this fact is so often ignored. Time after time I witnessed families concerns or interests - based on a history of love, care and attachment - swept aside by senior care professionals and clinicians as either insignificant, irrelevant or outdated; I lost count of the times that senior nurses described families involvement as "stupid", "idiotic", "irrelevant" or even "insane".

There is a distinction to be drawn here that is nuanced but nonetheless critical. Families concerns may be misguided and their requests often detrimental of the holistic care of a service user but their motivations are based on a justified and significant grounding of care and love – while it may not be possible for clinicians and carers to comply with each request a dismissal of them wholesale that fails to account for the underlying motivation of attachment is a disservice to the those families and a basic failure to account for the service user as part of a unit – the family - that transcends the ultimately transient nature of care, whatever its duration and depth.

The basis of this dismissal of families is systemic as on endless training courses I was party to being told that *we* knew service users *far better than families did*. This always struck me as presumptuous to the point of arrogance and bred the culture outlined above of "families know nothing".

From hospitals to nursing homes this culture is ingrained and, like a rotting fish, the decay comes from the head; while CW's are individually responsible at the level of care provided, and thus have an obligation to take into account families feelings, emotions and requests as they encounter them, the agenda as it is set is the responsibility of service or home managers and senior care staff and clinicians. It is wholly wrong that families are treated in such a shabby fashion by such senior staff – it diminishes care and it destroys relationships.

No factor is more damaging in care relationships than this arbitrary disregard of family's feelings and is one the major contributors to poor care, I'll state that in capitals to make the point. **ONE OF THE KEY ELEMENTS OF FAILING CARE IS TO DISMISS THE CONCERNS AND FEELINGS OF FAMILIES SUMMARILY** (see conclusions).

The failure in treating families as if they have nothing useful or important to say is key in failing care - as to disregard family's' concerns is to also diminish the service user as a real and whole person; families are critical to understanding the continuous process of living as it is manifested through the service user as an individual rather than one among many, and that the individual, *this individual*, has memories, hopes, desires and a full interior life – no matter how psychologically chaotic. Once a service user ceases to be an individual they also cease to "matter" in care terms, they become merely a collection of tasks orientated round a barely understood being and nothing more, this is how care fails.

All these issues I came up against in my time in care and presented some of the most intractable and unsettling problems that I encountered. There are no answers to all of the questions posed above, and while the maxim of not pleasing all of the people all of the time is true in life, in care it is more a case of pleasing none of the people none of the time; for care as it is structured at present means that time occupied caring for one service user is care denied to another. These issues are less profound, I found, in supported living but are compounded in residential care for elderly where staffing ratios are thin to threadbare and where the atmosphere and pressures are more like battery farming than care, but more on this later.

Retuning to my early days in care, the first relationship issues I ran up against were with the parents of "Service User B" - and through this lens I learnt much about just where a new CA like myself stood in respect of family and care relationships.

"Service User B" themselves had quite a high degree of functioning, although and they were able to live a relatively full life but that "full life" was trammelled by both their mental issues and the need for constant one-to-one care - as they were unable to function safely without support. However supporting "Service User B" was not a particularly arduous or intensive task as it mainly involved prompting to meet personal care – encouragement and supervision to see they washed properly and were bathed every other day – the preparing of meals and accompanying them on their many social outings which filled most days. In many respects this care was less intensive than meeting "Service User A's" care needs as conversation was limited and the care relationship was more distant, added to which the activities made most days fly by and threw up many different interesting and stimulating encounters.

The aspect of supported living that makes it so rewarding for many CW's is that they are drawn into the world of the service user as they make contact with the outside world. I was stunned by many of the activities that "Service User B" attended that were run voluntarily or with minimum government of Local Authority funding that added to the richness and sense of fulfilment in many service users lives.

Music clubs, art classes, coffee mornings, sports events, cultural involvement; nearly all these were supported by an array of the most unlikely types of people – bright-eyed pensioners, middle-aged men who looked more likely to be seen in the middle of a barroom brawl than supported living activities, fresh-faced youth workers filled with enthusiasm, housewives with busy families and lives of their own who still crammed in time and application to run activities in draughty church halls or community centres. All these were key in making "Service User B's" life much fuller and meaningful than it would have been otherwise; it was the difference between being "cared for" and being "supported".

There was nothing more enjoyable for me than to see a normally introverted and diffident "Service User B" drawn into social activities that would have been, on a purely diagnostic basis, considered beyond their capacities. Selfishly too I enjoyed talking to these volunteers and community workers and hearing how they gave with generosity their time without evidence of ego or self-satisfaction. In fact it was a stand-out feature of nearly all of them that they considered what they were doing as nothing more than being part of a society that is focussed on giving rather than taking.

The irony was, and I will try to avoid "soap-boxing" here, that many of these activities took place in the most run-down places in the city where I worked, places where wise folk didn't stray after dark, and were run by the supposedly indolent. A good minority of the volunteers I met were either unemployed or themselves on benefits associated with longstanding health issues, some of them were alumni of the very classes they now ran, who had been supported by other volunteers through their development and now were able to provide a service to others as it had been provided for them; other volunteers were simply using their time out of work to do something other than sitting at home watching the walls peel.

It also was a sad fact that just as I was just moving into working in supported living the minimum funding on which many of these groups depended – for rent, tea and coffee, for provision of materials and administrative details (all volunteers that have contact with potentially vulnerable people have to have DBS clearance) – was being cut by central government and Local Authorities. These are no headline services, they are the Cinderella's of the much maligned "care in the community", they do not fulfil critical services, they do not deal with acute care but they are the grease in the gears that keep many people's lives on track, or give them purpose or even meaning, they do not attract petitions or impassioned advocates when they are under threat, instead they melt away unnoticed by most, un-mourned except by the voiceless.

The lamentable point is that the benefits of these ad-hoc services are not accounted in the cost-benefit analysis of care for those with learning or physical challenges; they are part of a once lively sector that made care in the UK work. They are under-reported and their significance hard to quantify – how can you put a cost, price or result on the fact that they have filled an afternoon for someone who otherwise would have spent it alone? How do you quantify the worth of something that, like for "Service User B" – who existed according to set and comforting routines – gave a day of the week import?

Like unpaid family carers they are ignored or ridiculed as the occupation of "do-gooders" (just when was it that "doing good" got a bad name?) and like families they are often the butt of professional carers and clinicians malicious jokes.

While "Service User B's" family were happy that their loved one had a rich and varied life other care workers joked at the expense of those that ran these services; that they were a "waste of time", that it was a chore taking "Service User B" to attend them, and that - according to them – on what basis I know not as "Service User B", while maybe not appearing to be enthused by the actual activity, nonetheless was at a loss when they could not attend - "Service User B" "did not like them" or was "bored" by them. Quite what to be more stunned at was hard to decide - whether it was the fact these other carers found these classes "boring" so imputed these feelings to "Service User B" or that they made sport with denigrating those that ran such activities. Either way it showed both a gross ignorance on their behalf and a total misunderstanding of both "Service User B's" preferences and the importance these activities played for them.

Anyway, for me, once I left supported living these occasions of outreach were one of the main things I missed.

I supported "Service User B" on these and other outings, sometimes relying on them to direct me to locations as, when I was cleared by my CRB, I was often not told how to get to a destination and only given an address and so supported "Service User B" blind on these excursions, or, if reasonably prepared, with a Google map print out clutched in my hand. I also fulfilled all the other care tasks I was given around "Service User B", and, as far as I knew, I did them at least satisfactorily. However on meeting for the first time "Service User B's" parents I was totally unprepared by their frigidity and cold regard. The first time they visited to take "Service User B" out on one of their Sunday afternoon excursions I had the feeling of being, as HG Wells put it -

"scrutinised and studied, perhaps almost as narrowly as a man with a microscope might scrutinise the transient creatures that swarm and multiply in a drop of water."

I was aware not only of being observed in such a clinical and antiseptic fashion but of something more, it was as if I was being assessed from a less than favourable starting position, as if they were expecting some default position of a screw-up.

As noted above, this regard of relatives to CW's in general, and of a new face in a care team in particular, comes with all the assorted baggage that nearly all "outsiders" of the care environment bring with them when coming into contact with CW's - they have a predisposition to see all but known, tried and trusted CW's in a less than flattering light because conceptions of CW's come pre-loaded by a hostile (justifiable in particular incidences, less so in its lazy generalities) the media. However - and I admit that I am sensitive to such things - there was a little more to it than that. For the first of many times during my work in care I became aware of what may be called the "class and educational gradient" that often is a major effect and defect in the relationships between families and carers.

This needs a little expansion, but much of it will be self-evident in reading between the lines of what I have already covered in Chapter 1.

Most CW's are drawn from the working population who have received a scanty or poor education, and, although I met many CW's who's potentialities had never been tested and so lay undiscovered but who displayed many of the skills and aptitudes to progress much further, whether in care or in another field, the majority of CW's could be said to have a poorer education than average.

As to class CW's are mainly drawn from what was formerly called "the working class" with all the connotations of the respective social skills, background and societal position that this term brings. As many relatives are, or consider themselves to be from a "superior" class this creates an instant barrier that leads to condescension, patronisation, suspicion and a fundamental lack of respect.

These two (pre or post) conceptions inevitably lead to the attitude - if not the actual vocalisation - that CW's are "just care workers". Never has a phrase been so loaded and tipped with pejoration, never one so calculated to render individuals a lumpen insignificant mass since the word pleb entered the English lexicon, and like the term pleb, CW's are regarded as being - to borrow a description from the writer James Kelman –

"servant[s], brutalised....they're never fully formed human beings, never particular people"

CW's are not so much, in the main, looked down upon, as sneered at; as a group they are seen as a necessary evil, an unwanted but unavoidable consequence of the residential care process. They are certainly never deemed to have an interior life, any intelligence or ever worthy or even capable of holding a conversation with.

It may be thought I am going too far with this representation of the "class and educational gradient" that separates CW's from relatives, and maybe I am, but only in the fact that I am encompassing *all* relatives in it - I have known some notable exceptions - however it remains true in the main, and by no means hyperbolic, to state that in almost every interaction I have had with relatives' has been from a position of a prejudged educationally, socially and, probably morally, inferior standpoint to them. Time after time after time in care work I have heard families refer to CW's trying very hard to do a very difficult job as being not only the hoary "just care workers" but also "those people" or even "idiots" – I kid you not.

To give an example - I was working with another female care worker on an evening when the extended family of a service user were in visiting. Both of us were serving tea to service users and we asked the family if they wished to have any refreshment, rather than an answer the son of the service user simply waved us away with the back of his hand as one might do to an irritating insect, while the wife of the service user simply looked us up and down with an languid and superior air then ostentatiously looked away. As myself and the other CA left the room clearly audible was the wife saying –

"Why should we want anything from those idiots? They're only care workers."

Another example is even more illuminating, and, for CW's, depressing.

My partner was travelling away to a training course with several other members of the same company who she only knew in passing as they worked in another office from her. One of these colleagues had recently had to place their mother in a residential care home. Having been asked how her mother was coping with the change this person them launched into a diatribe about how those charged with her mother's care were "lower class", "lazy" "stupid" and worst of all "just scum". No particular incident incited this brutal analysis, no untoward actions or incidents, no failings, no problems; it was just someone saying how they regarded their mother's carers. Several of the others travelling with this individual – despite having no experience of seeing CW's in action - joined in on this general dishing of care workers, churning out preconceptions over CW's being negligent abusers, stupid and unable "to do anything else but work in care." My partner listened to all this patiently until the vitriol had stopped dripping before stating that I, her partner, worked in care. After the initial leaden silence that always appears like a bullet coloured cloud when people know that they have dropped something of a bollock all these colleagues then fell over themselves to say that obviously I was "one of the good carers" and they were not referring to me but to "the others" in care who were "stupid", "lazy", and "scum".

The fact that I was not surprised by this dishing did not prevent me from still being nettled; and I wasn't sure what nettled me more – the fact that these individuals considered themselves morally, socially and intellectually superior than unknown CW's (if this had been a collection of the apocryphal "rocket scientists" I maybe could have understood, however as they all worked in the cosmetics industry - one hardly free from negative preconceptions itself - it was somewhat too much of an irony to swallow without choking) or that fact that they had lashed out at CW's with no reason then elevated me above the Nietzschean "herd" of CW's with only the foundation that they vaguely knew my partner. Just as their bitter criticisms of CW's in general were misplaced so was their hasty "exoneration" of myself in the particular – as far as they knew I could have been the laziest, most stupid, low class scum imaginable.

Both these examples point to how families often view CW's with no prior evidence. Although "Service User B's" parents did not partake in such incredibly bad behaviour, their regard to me - while being infinitely more subtle – and middle class - was nonetheless just as penetrating.

It is an uncomfortable sensation to be looked down upon while trying to maintain a conversation with someone you have just met, and especially difficult for me as I felt myself grinding my back teeth and with the terrible itch to ask if they wouldn't mind looking at me as if I was a fucking person instead of a fucking "thing" that had been carried onto care on the bottom of a shoe; indeed the only thing between me and making this career ending outburst (and perhaps proving their preconceptions with my descent into the Anglo-Saxon) was the avalanche of ironies that crashed upon me.

First among these was my background, which was solidly middle class; indeed if we were to compare social backgrounds I could have stood on a more than equitable footing - so while I was being visually patronised they had no idea that they were basing their scrutiny on sandy foundations. Then there was my education - which my own parents had pushed me through and which I had continued on an ad hoc basis since the end of formal education as I found I enjoyed the process of learning in and for itself; again they could not have known this as they made the spurious link between the job and my intelligence.

Perhaps most pertinent of all these ironies though, and one which remained through all the dishings I got as a CA, was that while they were regarding me as something less than totally human they were at the same time entrusting the care of their son to me. It seemed less than politic to regard such an important cog in the wheel of care as the CW's who supported their son with such cold indifference. And this is the paradox that families find themselves falling into – they may regard

CW's a stupid, lazy, ignorant and base yet they are there to provide an essential service to those they love. Rarely can it be that such essential workers are seen as such disposable objects, yet that is the lot of CW's; abused, ignored, regarded with suspicion or open hostility we are still expected to provide the best of care all of the time.

The self-defeating nature of the low esteem in which CW's are held sometimes spills over into actual care – and how could it be otherwise taking into account the above. The constant diminution of the status of CW's inevitably leads to de-motivation of care staff, and what is worse is that the attitude of particular families often has direct and baleful effects on the associated service user. I have seen CW's be reluctant to give the best of care to service users because of the attitude of the family. This is of course inexcusable professionally, yet, on a human level, it is understandable – to be totally run down by a family leaves CW's short of the compassionate element of care, to know that whatever you do, however well you care for a service user, no matter how far you go in supporting that service user, that their family will still regard you as scum often leads to CW's living down to their reputation – if they are to be despised and humiliated despite doing a good job, the reasoning follows that they might as well be despised and humiliated for doing a bad one.

The truth that families have to know is that if they treat CW's as human beings and with a modicum of respect and politeness that would be due to any other individual then most CW's will break their backs in the care of their loved one. If however they continue to regard CW's as nothing more than industrial waste then they are partly responsible if care falls short.

Once more I emphasise *this should not be the case*. Carers should always do the best job possible regardless of their personal feelings; however it would also be untruthful of me to say that the care of an individual service user will never suffer due to the adverse attitude of a family.

Over the several months I supported "Service User B" the attitude of the parents did not differ, in fact if anything they came to regard me even more coldly. I am open to the thought that they believed – true or not – that I was not an adequate CA for their son; I am also open to the thought that they may have had some grounds; all I can say is that on every day I supported "Service User B" I did my best for them and could not recall any time, day or situation that I walked away from feeling that I provided less than the best care I could.

I was to learn later that even if I had performed wonders, been a cross between a modern day Anne Sullivan (Helen Keller's dedicated tutor) and Professor Brian Cox I still would not have been held in much regard by "Service User B's" parents, such matters though were out of my control and which I will come in the next chapter, however there was one incident that seemed the apotheosis of our tenuous relationship that is worth relating here.

One evening myself, the team leader, "Service User B" and their parents were gathered in the lounge and I passed an innocent and, as it happens accurate, observation about "Service User B"; it was neither controversial or provocative and was made in all good faith – that though was no excuse for a stupid intervention in which I was wholly at fault. No sooner had the words escaped my mouth then an icy silence descended. This was followed by a sharp comment from the father –

"We don't talk about such things in front of ["Service User B's" name]"

Realising I had made a crashing error I shut up but could feel the blood rising to my cheeks so I made my excuses and left the conversation. Later I was given the hook by my team leader and given a bollocking about discussing "prohibited" matters in front of "Service User B". I duly apologised and asked if I should apologise to the family and was brusquely told that would not be necessary – just why would transpire shortly – and that the matter should be left to rest there. The team leader turned his back on me and I was duly chastened. However I was still rather perplexed as to the gravity of my offense - what I had let slip was something that "Service User B" well knew and, although an uncomfortable fact, it was one they had assimilated themselves and their lifestyle to, further they also had the capacity to understand the context in which my unfortunate comment was made – in passing and signifying nothing more than part of a conversation – and was only an acknowledgement of the truth that "Service User B" accepted – it seemed rather than making "Service

User B" uncomfortable he actually appeared as if he was untroubled by my inapt comment; further it seemed it was the family who preferred this subject not to be raised, and that in itself had been my major crime.

Here was a distillation of everything I have discussed above – as it appeared to me. The lack of control that families feel leading to its vigorous reassertion through censuring me, the reductive notion that "Service User B" needed to "protected" from truths as if they were still a child, the dim view of CW's and their competence, and the carrying over of petty and vindictive disputes between CW's into the relative-carer relationship – something we shall cover next. All these coalesced around me at that point in one dazzling and humiliating moment.

Despite knowing all this I still I could not, and indeed still cannot, shake the image of "Service User B's" father as he had slapped me down – a look of sharp eyed flashing anger coupled with a strangely happy, or, more accurately, contented look that swiftly overtook this ire – it was as if I had confirmed something he had long suspected, like a man who shoots his toes off just to prove to another that a gun was loaded. Call me paranoid but I believe all his preconceptions were waiting for this moment, had been waiting for me to make a mistake – granted a foolish and ill-conceived one - and so vindicate his feelings that here was just another lower class idiot making an arse of a simple job.

Maybe he was right.

Either way not only were things soon taken out of my hands, but half the service too.

CHAPTER IX

Care Is A Racket

If "Service User B's" family as I knew them were unhappy with the care provided by myself and the service in general I could not say that these concerns were unfounded, at least as far as the service was concerned.

Right from the very beginning of my working in the particular service, and most especially working with "Service User A" and "Service User B" I could not for the life of me understand how the two of them had been housed together. "Service User A" was volatile, solitary, badly socially adjusted, enjoyed watching the TV and his collection of DVD's, abhorred any conversation that he wasn't part of, was lacking in any inclination unprompted to maintain personal hygiene, was destructive of property, had no family – and was painfully aware of the fact – and most seriously, he was a very intimidating presence whether he was in a room or not – he projected potential menace even from a distance.

"Service User B" however was – even taking into account his disabilities – quite well socially adjusted, quiet, reserved, well-mannered, scrupulously clean, took care of all that belonged to him, enjoyed clubs and groups, was placid and – critically - very fearful.

The consequence of this housing of such two disparate characters together was awful for "Service User B" who did not dare to venture into communal areas while "Service User A" was there, and if he did – as he had to for meals – he spent as little time there as possible – bolting his food down with one weather eye on what "Service User A" was doing, or, even if "Service User A" was absent, upstairs in his room,

"Service User B" ate with a cocked ear toward the sounds of movement and the potential materialisation of "Service User A".

"Service User B" dared not watch TV in the lounge and instead spent his waking hours, when he wasn't out at his many activities – shut away in a downstairs room that had been designated a "safe place" and nearly always with a support worker sat with him. At night he locked his door in case "Service User A" ever ventured into his room (he never did but that was beside the point) and on coming out of his room he would almost comically – if it weren't so serious – look left and right across the landing several times before he emerged. To say all this reduced the quality of "Service User B's" domestic life would be a chronic understatement, the psychological cost unfathomable.

When "Service User A" was away on holidays "Service User B" was almost a completely different person. They sat and watched TV in the lounge, ate leisurely meals, didn't lock their door at night and generally made the house what it should have been – a home.

Therefore a more ill-matched pair would be hard to find, yet here they were daily living cheek by jowl with each other. It seemed no one quite had the answer why the two were housed together.

In the absence of information I assumed that "Service User A" had been resident already in the house and that "Service User B" had been housed with him because there was space for two occupants and that the general environment – a small house, as opposed to one of the large communal housing blocks – was domestically suited to them.

Having been housed together as a matter of necessity, I assumed, even when additional spaces became free in other similar houses - and with other service users much more temperamentally suited to "Service User B" - the upheaval of moving "Service User B" and all their considerable possessions was thought to be too much disruption for someone so tied to habit, and so the failed "experiment" of housing "Service User B" with "Service User A" was therefore continued regardless of its unsatisfactory nature.

So it came as news to me that actually "Service User B" had been resident in the house a good 8 months *before* "Service User A" arrived. Worse, nearly every stick of furniture in the house – from the armchairs to the dining table – and every piece of crockery and cutlery, were "Service User B's" even though now his life was so inhibited that he could not avail himself of the use of hardly any of it.

More than this the service provider I worked for, and the service manager who ultimately controlled housing allocations, were well aware of "Service User A's" behavioural issues when they accepted him *and yet still housed him with "Service User B"*.

There followed – predictably - an extremely volatile period after "Service User A" moved in - where his adjustment from secure hospital to supported living had been horrendous. He had been violent, aggressive, abusive, and destructive and almost beyond any civilised control. During this time "Service User A" had smashed two TV's to smithereens, destroyed furniture - "Service User B's" furniture - put two windows out - one downstairs and one upstairs - resulting in them having to be replaced by reinforced Perspex ones - had assaulted care staff, and ultimately he had physically assaulted "Service User B" resulting in the police being called and "Service User A" spending the night in the cells of the local nick. He was only saved from a potential return to the secure hospital by the munificence of "Service User B's" family who, seeing "Service User A" as a troubled soul, had no wish to either press charges or to see him retuned to whence he came.

This generosity though did not stretch to seeing their son become a potential Aunt Sally, and immediately noticing the effect of the housing of "Service User A" with their son they had asked that he be moved as a matter of urgency. Apparently the service manager had agreed and told them that "Service User B" would be moved as soon as possible - once a suitable vacant place had opened up. This was almost 2 years ago and there had not only been no move and no movement toward it, but there had been no sign of the service manager having any real intention of moving "Service User B" as in the time since several suitable places had opened up but no action had been taken; this was even after countless meetings the family had had with the service manager and frequent written complaints. Even after they had taken the issue up with the regional manager of the service provider it resulted in nothing more than more promises and even less action.

This shed a little light on the situation, if complaints had been restricted to the service manager alone then lack of movement could have been put down to intransigence, incompetence or laziness, however having elevated it more senior management the issue appeared to be something more fundamental than issues of competence or will – it appeared to be a corporate decision taken at the corporate level. It took one shift with a hugely experienced fellow CA, new to the service, for the whole situation to become crystallised.

It was money.

In brief "Service User A" had plenty and "Service User B" very little.

To expand a little –"Service User A" - having no family and having been virtually incarcerated for much of his life - had been in receipt of benefits that he had never had cause to spend, so had built up a bank balance in the low thousands. Added to this, in the local Primary Care Trust's (PCT's) keenness to reduce the numbers of long term psychiatric in-patients to meet targets and cut costs, when "Service User A" was turned out into the community they attracted the largest package of benefits possible from the relevant Local Authority *and* the PCT – in short they were supported with a relative wall of money. As "Service User A"'s tastes were simple, and as he never ventured out apart from trips in CW's cars and twice yearly holidays, he never spent all of what kept coming in every month, so his bank balance kept increasing.

"Service User B" however came from a family that on paper was quite well off financially; as he had never spent long periods in residential psychiatric care, had been looked after by his parents in the family home and had been educated privately, his social income was low. Combined with this was his rich and full social life – all of which cost money. The result was that he had less than a couple of hundred in the bank. What "Service User B" did have though was plenty of material possessions.

The CA I was working with told me it was standard practice in care to match someone with large cash reserves with someone with fewer financial resources. The idea, and it proved to work generally, was that through sharing funds for food and other communally consumed items – cleaning materials and washing power and such like – more could be provided for less real cost to a "poorer" service user than if they either lived alone or with another similarly cash strapped service user.

In the particular case of "Service User A" and "Service User B" it ran like this - "Service User A" contributed more to the weekly shop than "Service User B" under the pretence of that fact that as he rarely ventured out and therefore he would use the lion's share of the weekly shop - ergo he paid in more. And there were other, more subtle ways, of using "Peter" - "Service User A" – to subsidise "Paul" "Service User B". If a hoover needed replacing it came out of "Service User A's" account. If a new washer was needed, likewise; and so on, the result – according to theory – was that the poorer party would be better off. It was only a happy circumstance that "Service User B" contributed by the use of his furniture that otherwise would have needed to be bought.

The result was the mismatched pairing, the cost, the heavily circumscribed life that "Service User B" was forced to live. It turned out that prior to "Service User A" appearing on the scene it was uncertain whether "Service User B" could be kept in the service while living alone, other potential house sharers were as impecunious as him so a cash cow was needed – enter "Service User A" and his bulging bank balance. The whole point of the exercise ultimately was to keep "Service User B" in the service, to keep the money rolling in.

The ugly fact that this exposes is that – **SERVICE USERS ARE PLACED NOT ACCORDING TO PERSONAL NEEDS BUT ACCORDING TO THE FINACIAL BENEFIT THIS WILL PRODUCE FOR THE SERVICE PROVIDER.** The case in point was the one I was involved in.

It has to be said here that the service provider I worked for was one of the less egregious in this respect and there have been cases where whole services have been set up solely for the purposes of generating income, not out of genuine need.

Let us turn back to the Winterbourne View Serious Case Review (SCR) – numerous excepts show just how much finance rather than care plays a role in care provision.

Winterbourne View, from the very outset, was a facility based on potential financial returns rather than genuine care need. The SCR notes –

"Winterbourne View Hospital made no reference to government policy in terms of developing local services for local citizens and closing long stay hospitals."

Further it was based on a feasibility study in which one of two of the primary terms of reference were -
"Current and future demand for Learning Disability services [and] The current supply configuration and future market opportunities"

Following the study Castlebeck Ltd (the "for profit" operating company of Winterbourne View) concluded that–

"Bristol and the surrounding area.......Provided a market opportunity to develop an Assessment and Treatment service"

The result of this was Castlebeck Ltd's Board took the view that rather than developing services within the community – of which there was a genuine need and which was part of government policy -

"That developing their own hospital would be more commercially viable."

Thus Winterbourne View. A classic case of commercial considerations trumping care needs.
Once Winterbourne View was open things got worse –

"Although Castlebeck Ltd's Management Review describes Winterbourne View Hospital as *one of the best performers within the group from a financial perspective* [my italics], it does not appear that the weekly charges were directly tied to the cost of the various components of assessment and/or treatment. Correspondence with Castlebeck Ltd in November 2011, confirmed Winterbourne View Hospital's turnover as £3.7m in 2010. Given that no amounts were returned to shareholders or management (apart from salaries) in that or any other year, Castlebeck Ltd was asked, on average, how much of the £3.5k charged per week, per patient, was spent on (a) patient activities (b) physical health care (c) psychiatric input (d) nursing staff (e) support workers (f) assessment and treatment (g) food and catering (h) heating and lighting (i) laundry and cleaning (j) maintenance and repair and (k) administration. *The company declined to answer because of the "commercial sensitivity" of such information* [my italics]".

Because of the financial success of Winterbourne View –

"The high levels of sickness with a number of disciplinary actions…5.3% in 2009 and 6.8% in 2010, did not invoke Board level/ Executive Team concern."

The implication is clear; Castlebeck did not want anything as troublesome as investigating and uncovering abuse to interfere with its cash cow.

The SCR made these conclusions clear -

"Business opportunism, which was not discouraged by NHS Commissioners, was associated with the development of Winterbourne View Hospital as [a] hospital for adults with learning disabilities in 2006."

Winterbourne View was not an aberration though, the Francis report into the Mid Staffordshire NHS Foundation Trust disaster (which it was, however slow in evolving, however un-dramatic on day to day terms) notes as one of its primary findings that –

"There was an unacceptable delay in addressing the issue of shortage of skilled nursing staff. There can be little doubt that the reason for the slow progress in the review, and the slowness of the Board to inject the necessary funds and a sense of real urgency into the process, was the priority given to ensuring that the Trust books were in order for the FT (Foundation Trust – a procedure where a hospital gains control over its own finances and receives a boost in direct government funding) application. The result was both to deprive the hospital of a proper level of nursing staff and provide a healthier picture of the situation of the financial health of the Trust than the true reality, healthy finances being material in the achievement of FT status. *While the system as a whole appeared to pay lip service to the need not to compromise services and their quality, it is remarkable how little attention was paid to the potential impact of proposed savings on quality and safety* [my italics]"

Stated much more baldly –

"The Trust prioritised its finances and its FT application over its quality of care, and failed to put patients at the centre of its work."

Once more financial considerations trumped care. These problems of varying severity can be found right through care. In care homes the situation is critical, as the Financial Times uncovered in its investigation into care homes in 2011
http://www.ft.com/cms/s/0/920fcd2c-8aca-11e0-b2f1-00144feab49a.html?siteedition=uk#axzz24q64GYMr
In summary it found that the private sector care provision was almost totally failing in the care *they were paid to provide by taxpayer's money* in favour of boosting profits. The full proof of this assertion can be seen in more detail in the investigation -

http://www.ft.com/cms/s/0/307bbd3e-8af5-11e0-b2f1-00144feab49a.html#axzz2hflpXvRx

And the shocking truth is that matters have got worse since the 2010 investigation of care by the FT - tightening fiscal conditions, government austerity measures that affect Local Authority funding and ill-considered business models predicated on the fact that there would always be a ready supply of service users bringing with them lavish packages of funding, have all contributed to cash crunches that have affected individual care.

For-profit companies predicated on making money off the back of the ill and the old face what is a central conflict of interest – that of making money against provision of good quality services that meet service user's needs. It doesn't take genius to work out that in a funding crunch which will be sacrificed first – earning a healthy wedge or looking after the service users and staff adequately?

Take a lucky guess.

The situation in care homes, which I shall come to later, is the most critical and the most disturbing in respect of care quality, all this leads to an unavoidable conclusion that is not only apparent from the empirical data but also from my direct experience and that is that **CARE QUALITY IS SUFFERING AND FREQUENTLY TOTALLY FAILING BECAUSE PRIVATE CARE PROVIDERS ARE PUTTING PROFIT AHEAD OF CARE IN THEIR PRIORITIES AND THIS SITUATION IS ONLY GOING TO GET WORSE IN TIME.**

This is the result of treating service users as commodities, when individuals are reduced to revenue streams and performers in the profit column of the profit and loss account; if placed on a balance sheet, service users would be termed an asset. This is where we are in care – the reduction of human beings to mere financial transactions; and here's the rub, if vulnerable people are treated as nothing more than revenue streams what message does this send to care workers? And if this overall free market based corporate guiding ethic is one that reduces humanity to a financial transaction – which it does - how are CW's meant to maintain the dignity and self-respect of service users when they see these same principals are totally lacking from the corporate calculation? For that matter, witnessing how indifferent service providers are to service users does this not bring on the recognition, alluded to many times above, of CW's true status and the "regard" they are held by their employers – that they too are commodities, infinitely replaceable units of production to be picked up or discarded at will. How then are they to maintain motivation in what is often a very difficult and stressful job? How are they to maintain their own self-respect and dignity in order that they should recognise the importance of delivering it in their care?

The answer is they try but fail; the demotivation crystallised in the awareness that are only protecting and serving a commodity cannot help but influence care principals detrimentally.

While the actors and malefactors in the in Winterbourne View and Mid Stafford hospital cases bare *absolute personal responsibility and accountability for their actions*, then at least some of – perhaps even a greater - burden of responsibility and accountability should have been shouldered at corporate level *in that the leadership that was given that placed financial imperatives ahead of even the minimum care*; in brief - the individuals committed the crimes and unforgivable failures of care, but the leaders of both organizations *set the conditions in which these crimes and failures where not just tolerated but allowed to flourish;* and yet while carers – rightly - went to jail and a host of nurses were disciplined no one at a corporate level received equal sanction; in fact they escaped scot free.

That both cases have not acted as a wake-up call to the fact that financial imperatives *still define how much of care is delivered* then it hard to see what will be. If anything positive can be said of the present it is that it is *infinitely better than the course we are charting to the future*, which, if we maintain it, will see many more Winterbourne View's, many more Mid Staffordshire hospitals.

The effect on "Service User B" was nothing so spectacular as the above cases, but it was a heavy and very real human cost. It was not the severity of the living restrictions that he faced but *the fact that there were any living restrictions placed on him at all* in what should have been an environment in which he felt free and safe.

The family of "Service User B" continued to agitate for a move but they now seemed resigned – a situation that suited the service provider as what they were counting on most was the "inertia" (a term explained below) of "Service User B" once they were "secured" by the service. The reluctance of most families to put their relative through the upheaval of a move to another service is one of the main levers that service providers have once they have a paying service user on their books.

The cold brutality of the calculation in this takes some fathoming and gives us a reality check on where we are with adult care services.

Service providers know that most of the service users they are "managing" depend on a rooted sense of home, a place in the world and a position of relative security. With most adult care service users they will have undergone at least one wrenching change – either the move out of the family home where they have previously been cared for, or out of another service that could no longer cater for them either by virtue of age or needs. Having often taken some time to adjust to their new environment and having undergone potentially several huge psychological costs, once an adult service user is in situ in a service most family members or care managers are reluctant to change this – this is what I mean by inertia.

In "Service User B's" case this was especially true - the service provider knew how important routine was to "Service User B", that they had adjusted after a short time of depression following the move to the service, and that now they had the full and rich social life - which I described above; therefore the calculation was that the family, no matter how they moaned and groaned, no matter how many meetings they organized and no matter how much noise they made, would not take the drastic step of removing them from the service. As the parents of "Service User B" were now quite elderly and were experiencing several health problems of their own it was probably felt at a corporate level that the "problem" of the parents would soon be "removed" in one way or another.

Again I ask anyone to consider how we have come to this state of affairs, that service providers – even relatively good ones – and the service I am discussing here is one of the better ones out there – have come to view people so indifferently that they feel they can put individuals into wholly unsuitable placements secure in the knowledge that because of the very challenges that lead them to needing care support in the first place will also infinitely reduce their chances of moving out of the service. How have we arrived at the point where an individual's difficulties in living should be viewed as a corporate business benefit and opportunity?

As the FT investigation into adult care services noted –

"Britain collectively does not care that much about its elderly, handicapped and mentally ill."

If though this calculation of inertia as applied "Service User B" was made it was about to go deeply awry, ironically due to the fact that care around "Service User B" was indeed "support worker led" and that the family's refusal to be fobbed off was stronger than their age and health would appear to dictate.

CHAPTER X

Moving Out

Always the last to know I got to hear that the "Alpha Dog" of the care team was leaving to go to another service about 2 weeks after everyone else appeared to be aware of it. There seemed no exact precipitating event other than the constant dissatisfaction with the service manager, the service and everything else that had been accumulating - all poached in bile. Yet to say I be would glad to see the back of him would be to be unfair – something I couldn't say for the other 2 members of the cabal who I couldn't care less if they fell under a bus. The Alpha Dog had in fact been very helpful to me in picking up the job, he genuinely seemed interested in seeing be get on and provided many insights and tips that proved indispensable, especially in supporting "Service User A" with whom he had a good rapport. So it was with mixed feelings I wished him well. 2 weeks later another of the cabal put in their notice, in this case I couldn't have been happier as this individual was the second most unpleasant person I have ever had the misfortune to stumble upon – the top spot with a bullet being occupied by an individual we shall meet later.

This Second Most Unpleasant Person I Have Ever Had The Misfortune To Stumble Upon™ and the female CA of the cabal had been provocateurs of some of the most uncomfortable times I had spent in the service. They had never concealed the contempt in which they held me and did everything they could to make me feel like an outsider, an interloper, a fool and a lazy one at that. No matter how hard I worked, no matter what initiative I made, no matter how helpful I tried to be there seemed no way of being accepted as either a fellow CA or even an individual. I dreaded shifts with them because they would make the job unremittingly unrewarding and, at times, unbearable. Further, I felt like I was constantly on trial, I could feel their eyes roving over everything I did, searching for shortcomings, defects or lapses, at no point could I relax or be at ease, so it is of little surprise I looked upon his departure with little else but relief and relish.

Equally unsurprising was his destination - the same place that the Alpha Dog had moved to, so it was no shock when, 2 weeks after he left the female member of the cabal put in their cards as well, once more, and inevitably, heading to the same place.

This is what makes such "cabals" that appear in care so difficult to explode, an Alpha Dog who moves is no longer the Alpha Dog without their sycophants, and the sycophants, without someone to suck off are just, well, you know where I'm heading with this simile; so it often happens that when a "leader" leaves the rest are compelled to follow.

It was only after they had gone that the damage they had done became apparent. All of them, including the Alpha Dog who had seemed so solicitous, had dished me to other CW's, managers, and, worst of all, to "Service User B's" family. I felt betrayed, and yet foolish to feel betrayed, as if I had been staring the fact in the face even though it had been unknown to me; I felt I should have expected it, prepared myself for it, defended myself from it, after all I knew the bitterness of their vitriol as it had been projected in my presence toward others so why should I have expected anything other than that same vitriol to be projected onto me in my absence?

A horrible feeling of disappointment pervaded me. Even though I knew I had done nothing but my best as I was able to on any given day I still felt I had failed somehow. This annoyed me even more, at heart it was a childish desire to please, to be well thought of, to be *liked*. At the same time I was deeply irritated that the opinion of people who, in the main, I had had little respect or time for should harm me from a distance like this, after all, I reasoned, they were off on their merry way without giving me another thought and yet here I was consumed with thoughts of anger, bitterness and vengefulness but with no object; I was unsettled to think that I was wounded because I had *allowed* myself to be wounded. This was the second worst personal experience I had in care, something only topped later, but for now the feelings were devastating enough. Still I was comforted by the fact that other CW's I went on to work with thought me a good carer who had potential, that I should not give the cabal any more time in my head and that nothing I had done was wrong.

I am aware and feel myself squirm inside to write these very emotions down. If I were being less than honest about myself I would say that long experience of work allowed me the shrug off such petty slanders and that I was above them. But in that I would be guilty of being honest about others and not myself, and if I cannot and will not be honest about myself then none of what I have to say in respect to care would have any of the import of truth – lying about, and to yourself unutterably alters any individual's capacity to tell the truth about anything.

There is also another purpose about being so embarrassingly frank about how I felt then - and that is to demonstrate the real cost that such cabals as form in care can do to other CW's. If I, at nearly 40, and having worked in some of the most tough jobs around, found the cost hard to bare in psychological terms, if I found my self-worth in my job so undermined and if I carried round the hurt for quite some time, what would be the effect on someone much younger or potentially much more vulnerable than myself?

As it transpired I was to see first-hand the effect of such psychological brutality on other CW's and less hardened than myself; I witnessed CW's virtually destroyed through whispering campaigns, sly digs and tale telling without foundation to management. I have seen perfectly good CW's undermined and potentially excellent carers leave the profession because of the very bitterness and almost wanton cruelty at work. Managers and senior staff have a duty to stamp on such practices because most care services have an anti-bullying policy, however this is more often honoured in the breech rather than the observance. Further, some managers – and I have known at least 3 – actively encourage this backbiting because if the workforce at odds with itself it cannot unify to tackle issues relating to management; in care divide and rule is *the* fundamental management principal. The best way at present to tackle such practices is for any worker on the end of them to ignore them as the cabal will soon move on to another target; but this is weak advice and of little use, it is further eroded by the fact that real damage can be caused by such maliciousness.

One manifestation of this was my deteriorated relations with "Service User B's" family. As already mentioned my relationship with them was less cordial from the start and was based from the outset by suspicion and an implicit lack of respect. With all the untruths told subsequently to them by the cabal the relationship went into something of a death spiral, every time I had cause to speak with them there was a barely concealed hostility in our conversations from their side. When the Team Leader announced that he too was leaving I sensed that something was afoot. Although his decision to leave was not connected with anything other than a personal desire to pursue his career elsewhere, the result of this personal decision precipitated action from the family. In the weeks leading up to his departure I was witness to several private conversations between him and the family in the staff room of the house with the door closed and low voices humming unintelligibly from within.

It is a sickening feeling to know that something is going down that will have an effect on you but which you know nothing in detail about or can do anything to influence or even be a party to. I privately sweated and fretted that these conversations could be inclusive of me and have a bearing on my future in care. Although that may sound egotistical and paranoid from without, from within care I had already, even at this early stage, attuned myself to the way that care works – if you think people are talking about you, you are probably right.

The last day of the Team Leader's day of work coincided with the breaking of the news that "Service User B" was leaving the service to go to another one in the next town to the north. In some ways I was totally unsurprised in others I was shocked. I was unsurprised because the family had been battling inaction in the service for over 2 years now to get "Service User B" into more appropriate accommodation, having been repeatedly fobbed off they had taken matters into their own hands.

Or not quite.

It happened that the decision to move "Service User B" - while being mooted within the family for some time - had been taken finally with the departure of the cabal, who had so poisoned the atmosphere around the new care team that was being assembled that they felt it was their only choice. I could not blame the family for this decision, in their shoes I would probably have done the same thing, the fact remains though that I believe that the care team that existed *after the cabal had left* including two young, enthusiastic, and receptive CW's and a new Team Leader - who was very experienced and empathetic to "Service User B's" situation - was, even excluding myself and any personal bias, *better than the one it replaced*. Before anyone thinks – he would say that wouldn't he – this conclusion was based on more than just antipathy to the Ancien Régime.

The two new CW's were bubbling with ideas and initiatives that previously would have been squashed by the cabal – making their customary decisions *for* "Service User A and B" without consulting them. They also benefitted from an atmosphere generated by the new Team Leader that gave them latitude in trying out new ideas without the fear that if they failed that they would be hammered by their fellow CW's. It was ironic then that "Service User B's" family had taken the decision now to move him.

A further irony was that service users were being moved out from their edge of town location by the service provider and into the deeper community anyway; it was therefore possible that over the next 12 to 18 months "Service User B" may have been placed in more suitable accommodation without moving him from the service. Yet it is not surprising that the family didn't wait around to see if this would actually be the fact – having being made so many false promises and subjects to illusions for so long.

While the above elements disappointed me but didn't surprise me. What did shock me was the fact that the departing Team Leader had been actively looking and finding a new service for "Service User B" – knowing he was leaving he had facilitated, and to some extent initiated the move.

Once more I felt more than a little betrayed, after all I had had good relations with the departing Team Leader who had been an excellent mentor and someone who appeared to have every confidence in my abilities as a carer; yet here he was appearing to undermine not just me but the whole care team.

Of course it was not all about me, or about the care team but there was something about the covert nature his on his part in facilitating the move that appeared underhand. On the other hand, and to be scrupulously fair, he was probably bound by confidentiality within the family unit who did not want the news of the move to break without all the ducks being in a line, so that when they finally broke the news it would be a fait accompli, so to some extent the "cock and dagger" routine was, to some way, understandable.

The above sentiments may appear rather egocentric – after all what really counted was that "Service User B" should be located in a service that best fit his needs and wants – most especially for his personal safety and liberty – however, as noted several times so far, CW's are not just carers but also humans, and it was thoroughly human of me to feel somehow a little betrayed by the covert nature of the planning and execution of the move, and the feeling that it inferred that not just the cabal, but also the departing Team Leader, had little faith in me or my abilities.

Again this reinforces the ontological nature of care for carers – at least good carers – caring is not just a job, it is part of your being; the care you give and the responses you get to that care go right to the heart of who you are, a failure in the care given – at whatever level, in whatever manner - also feels like a failure of your being, it feels like a personal flaw, a comment, a judgement – a *moral judgement* – on yourself – that you are not *a caring being*.

At least this is what a failure of care felt like to me, I was not just a carer between the hours of *being* on-shift I was a carer with my *whole being*.

Clinical professionals, even perhaps certain laypeople, may observe that these feelings do show a flaw in me – a failure not of care but of being able to distance myself from my job, a propensity to get too involved and not observe the distance that Richard Senett commented on - and which I quoted earlier in the book – that is the distance between yourself and those you care for. I would accept this, and yet it was the only way I came to understand care, the only way I felt I could be a good carer, I could not switch care on and off, care to me was a 360 occupation.

From this angle it perhaps is understandable why the removal, of "Service User B" - and the way it was carried out - had such a devastating effect on me.

What followed – although blackly comical verging on the farcical – showed that far from providing less than optimal care, the new care team in fact performed in ways that only made the potential loss of opportunity for "Service User B" from not staying with service more profound.

Having given notice to quit our care team spent at least 3 days a week taking "Service User B" along to the new service for acclimatisation visits in which he would left with the care team at the new service for 5 or 6 hours while the CW's who had escorted him there kicked around the small market town where the new service was based.

This didn't just make us feel like turkeys presented with an advent calendar, it was like being asked to form an orderly queue for the chopping block while Bing Crosby sang White Christmas.

As touched on above, it was no better testament to our care team that they performed this task with good humour and dedication, all the while knowing that they were soon to be caught in a personnel squeeze. For the fact remained though that our service had been staffed for the care of 2 service users, now there was only one. Despite me being not so hot on maths, even to my calculations the figures didn't add up, not with visible cuts being made in other areas by the service provider.

A further complication was the fact that the service provider was moving its regional base to a town 15 miles away and that service users already being moved out in to the community were being housed, in the main, close to this new centre of operations for logistical reasons. This meant that I would more likely be facing a 30 mile round trip on busy roads throughout the year, in all weathers, at all times of the day on either small and treacherous roads or large and very busy ones if "Service User B" were to be similarly housed.

Concerned about my position now "Service User B" had left I approached my line manager. I said that if I was to be offered shifts at the new location it would be very difficult for me to get there. The answer they gave took me aback

"You'll have to buy a car then won't you."

This was the among the closest expressions - in unvarnished truth - of how CW's are viewed and treated by service providers, the low stock in which they are held and the notion that they are easily replaceable.

The fact is, and this is an key point that needs highlighting –

NEARLY ALL SERVICE PROVIDERS COULDN'T CARE LESS ABOUT THEIR EMPLOYEES

But they should, the reduction of individuals to so many cogs n a machine to be tossed away if they no longer fit into any alteration in the machine's purpose or direction amounts in many cases to a take it or leave it option just like the one I faced above.

It should come as no surprise then that faced with this choice that many CW's chose to leave it.

But changes in service locations or priorities that prove impossible for CW's to meet are only a part of why so many are effectively forced out of the profession, other factors play a part, as will become even more clear in the following.

CHAPTER XI

Moving On (Or Trying To)

"A lack of career progression is a significant factor in the difficulty the sector has retaining staff. Turnover rates are high, standing at 19% a year in Care Homes and 30% a year in Domiciliary Care. Dr Shereen Hussein points out that the turnover rate for social care is 'considerably higher than that of other work sectors in the UK, standing at an average of 15.7%'. Research by the National Care Forum demonstrates the reasons for leaving; most significant are 'Personal Reasons', 'Resignation', 'Dismissal' and 'Competition from other employers'. *Deteriorating working conditions, poor training, as well as limited opportunities for progression, is encouraging workers to change between Care Providers, or leave the sector altogether* [My Italics].

The Kingsmill Review: Taking Care.

http://www.yourbritain.org.uk/uploads/editor/files/ The_Kingsmill_Review_-_Taking_Care_-_Final_2.pdf

Care has an employment problem; the above quote from the Kingsmill Review highlights many of the issues that I will come back to, but the eye-catching primary point are the figures highlighting the shocking attrition rates in care – nearly a one fifth of the workforce in residential care and nearly a third of domcillary carers either leave a particular service – common – or drop out of the care workforce completely – more often than you would imagine. This creates an enormous "churn" in a sector where service users often rely on a familiar face, a care relationship that has been nurtured and is co-operative in nature and predicable routines.

The Cavendish Review into Healthcare Assistants and Care Workers backs up these figures and takes a measure of their effect -

"Perhaps as a result of low pay, attrition rates are very high: around 19% a year in care homes and between 20.9% and 30% in domiciliary care. *This creates enormous problems of continuity for people who use these services* [My Italics]. It also creates significant recruitment costs for employers. Even during the current economic downturn, care homes and domiciliary care agencies are struggling to recruit.

When economic growth resumes, there will be an urgent need for greater workforce planning."

In care the normal rules in employment don't apply. The idea that, should the service change, or its priorities alter, that a "viable" or "reasonable" alternative be offered in order to facilitate an employee to remain in employment is an anathema.

Such was my experience - that I should be told that I would have to incur the cost of buying, insuring and running a car just to hold onto my job and on a wage that was less than substantial was a choice that was no choice.

(The irony of the occluded nature of choice as provided to CW's when choice is such a key component of care for service users should not go uncommented upon here.)

This wasn't a case of intransigence on my part either, I had taken the job knowing the difficulties in transport in getting to and from the present site – a 12 mile round trip that I had undertaken every day through one of the longest and bitterest winters in recent memory and never once had I been late or failed to show; neither had I refused outright to undertake a potentially longer journey, I had only approached my line manager to see if there may be a vacancy at a site closer to home. The reply that I got disillusioned me to the point of resignation as it displayed how expendable I was to any plans the service had going forward.

The other side of the coin made no apparent sense either. The service provider had offered me a job when I had no experience or recent credible work history – and for that I was truly grateful – but now having spent the time in shadowing me into caring for "Service Users A and B" – a period where they were effectively paying for two people to do the same job – and the cost of training me on top of that, they were now, seemingly, willing to discard me without any hesitation, this was, by all appearances, a huge waste of money and human capital.

Here was the dilemma that many CW's find themselves in, being given alternatives that are no alternative and feeling like they are almost an annoyance to be kept employed when all they have done is inquire into possibilities that will keep them in work - including being flexible to the point of breaking. Service providers here hold all the cards in a loaded deck, using the fear of redundancy they can make CW's play ball even knowing they are going to lose. Anyone with a vague knowledge of employment law knows that if work cannot be offered at a site where it had been up until now employers have a duty to see that they offer a viable alternative. The question here though is what is a "viable alternative?"

In care - built into most contracts is the caveat that runs something like –

"Flexibility will be required to meet the needs of the service which may include changes to any work location."

This is a catch all that absolves all service providers from having to sincerely provide "viable alternatives".

Seemingly in the situation I had found myself in the "viable alternative" was the purchase of a car and the onus of running costs; in other cases I have known CW's be forced to travel for up to two hours to and from the "viable alternative". Some would say that 2 hour journeys are not unacceptable as many commuters undertake this and more. What makes CW's different though is that they almost always work 12 hour shifts - usually on an 8 to 8 basis (8am to 8pm day shift, 8pm to 8am night shift) - so CW's faced with a couple of hours travelling time would have to leave their house at 6 (a.m. or p.m. it matters little) to get to work for 8 then take until 10 to even get home, after which they are expected to do this all over again, oftentimes for multiple shifts in a row. In this light it is not surprising that many CW's arrive at work unprepared or unable to provide support and care of the best quality.

Further to this are the costs of such transport; while a worker on the average wage may be able to absorb those costs, someone on minimum wage will find it much more financially debilitating to do so, it is not uncommon to find CW's working at least one shift a week that simply goes on travelling costs getting to and from their job, this makes the marginal cost of being in work negligible.

Of course employers couldn't care less about this - so long as a worker turns up on time where they are supposed to - but they should; the added travelling time and costs all have an effect on worker performance and contribute to staff burn-out, fatigue, stress and sickness. At present though it is simpler just to ride roughshod over workers and damn the uncalculated costs because employers also know that they dealing with a largely unsophisticated workforce who either will not know or understand the few rights they do have or else they come from overseas where worker protection is even less than that of the UK and so expect little better than the shoddy treatment they receive (the role of migrant workers and their impact on care will be discussed later as it is of no little importance).

Employers rely on this lack of knowledge to exploit their workforce and they also have a powerful aid in preventing their workers from gaining an awareness that they cannot be treated like human waste. The one or two occasions where I have known individuals able and willing to make a stand quickly find themselves side-lined, snowed under with fictitious complaints and, one way or another, out of a job. The worst sanction though is that care, although sprawling, is quite a closed community in many respects, and a reputation for being a "troublemaker" spreads faster than a wildfire; workers who have so gained a reputation simply by standing up for the rights can find their options for further employment in care rapidly diminishing.

All these facts made me pause at the thought of any confrontation and instead made me look at my career options, it also made me think about what aspect of care really stimulated me.

One of the area I was interested in working in was the dementia field.

Here another dose of honesty is needed, although I was drawn to dementia care by the interest I had in the many forms of dementia, it treatments, and how people with dementia are cared for, there also was also some pragmatism to go along with my idealism. I recognised that supported living was moving more and more over to the domiciliary model of care – that is caring for individuals in their own homes - therefore I knew the opportunities for full time care based on one site and with certain individuals was not going to last.

At heart the worthy aspect of supporting people out of a residential setting is good practice, humane, responsive and sound both morally and socially, but it would be a mistake to think that these are the reasons that it is favoured – the one reason why domiciliary care is the model of the future is that it is cheap. CW's in this field are some of the most belaboured and exploited in a care sector that already functions fundamentally on exploitation; they are often employed on zero hour contracts, have multiple appointments throughout the day in which no care worthy of the word is possible and usually not paid for their travelling time or costs.

This was not a world I wanted to be employed in either for the intangible rewards of the job – caring itself – or on terms that would effectively have made me pay to work.

Dementia care though is different. While those with the creeping onset of dementia can be cared for in their own homes, with the progress of their illness it soon becomes impossible to maintain this and so are moved into 24 hour care – that being the care home sector. The increase in life expectancy and the illnesses that are attendant on old age means that care for this group of people will always be needed so a move into dementia care could also be considered a decent career move.

(The residential 24 hour care model or what may be called the Nursing Home Trap will be discussed at much greater length later).

While the role of continuity of employment in my thinking may smack of cold calculation – and I admit it is less than edifying, but I never said I was an angel - this is to place care in vacuum outside of modern work practices. In many ways what I was doing was broadening my experience in care and extending my knowledge – nothing wrong there; in fact reams of literature are devoted to telling workers how they must keep improving and shaping their skills in order to stay relevant to the jobs market - and yet, because care is judged differently – as if on some higher moral plain - some people may view this logical approach as being disengaged from, and antithical to, a caring attitude. *It is true* that care is different but CW's are not treated differently, not elevated to the status and social standing of, say, nurses; in fact they are dealt with worse in employment terms than almost any other workers in any other field – as noted above. With this being the case no sensible care worker can view their job in any other terms than of continuity of employment.

As it happened the service provider I worked for had a dedicated dementia unit and I had been interested in working there for some time. I asked my team leader if she could introduce me to the manager of that unit with a view to working some shifts there, she duly did so and after talking to the unit manager I arranged to work on some of my days off-shift at my usual service.

Although the experiences of helping to care for my father had prepared me for the worst, last and most desperate stabs that life makes under the remorseless pressure of impending mortality, and the emotional cost of seeing the slow seeping of first personality, then self, then awareness and finally of breath itself, I was, unknown to me as I made my way along to the dementia unit on my first shift there, only experienced in the ways in which a mind and presence is ultimately betrayed by a body grown unruly then ungovernable; what I was not prepared for was the alternate betrayal of a body by the mind.

Avoiding any simple Cartesian implication of which is the most pre-eminent in the making - or the downfall - of a human being I suppose I had always seen death either in the terms of advancing years - as the decay in both mind and body equally; the common – and false – notion of a winding down of the life clock with the body and mind taking equal parts of the damage; or else in seeing the body only bearing the brunt of some physical imposition of disease or fatal calamity; it had never really been brought home to me that the mind could bring the temple down on itself by becoming so disordered that it could no longer master, control or maintain the physical vessel it was dependent on. This was the world I entered when I knocked on the door of the unit to gain access to the key pad secured building – reminding me the lock was there not to keep *others out* but to keep *individuals in.*

In the moment it took for the door to be answered and a suspicious night carer - unaware of my picking up a shift – warily and watchfully allowing me in - I momentarily panicked about the wisdom of this leap into the unknown. Once inside I rapidly became aware of the fact that I was now in a completely different care world to that which I had been introduced to up until now.

In many ways it was shocking, yet in others it was enlightening and rewarding beyond compare.

The dementia unit itself, although in many ways of a classic sort – smells of soap and linen riding over disinfectant and leaving the faintest stain of decay to cling to its antiseptic coattails - was in fact a specialist facility for those with early onset dementia (usually defined as those under 65 years of age), specifically those with learning disabilities who had been deemed to have grown too "aged" for the conventional supported living service. It also catered for those with Down's Syndrome who minds were failing them, as such no-one there was above the age of 65.

[N.B. It is a fact little known until the modern era that people affected by Down's Syndrome are particularly prone to early onset dementia. This has only recently become apparent as better treatment - medical and psychological - and better care has now extended the life expectancy of individuals with Down's Syndrome - since the 1980s their life expectancy has doubled and many now live into their 60s. The Alzheimer's Association in the US points out that -

"Studies suggest that more than 75 percent of those with Down's Syndrome aged 65 and older have Alzheimer's disease, nearly 6 times the percentage of people in this age group who do not have Down's Syndrome."

And -

"Autopsy studies show that by age 40, the brains of almost all individuals with Down's Syndrome have significant levels of plaques and tangles, abnormal protein deposits considered Alzheimer's hallmarks."

For the estimated 60,000 people with Down's Syndrome in the UK, and those that care for them, this is a growing concern]

Many of the service users there were in the advanced stage of dementia and needed support with all their care needs, this meant total care. Almost all of them had lost all mobility and so had to be moved by hoist from bed to chairs, chairs to bath or shower, from shower back to chairs and then back onto the bed to be dressed before being returned to their chairs ready for the day.

The chairs used were not conventional wheelchairs but bespoke ones measured and built for the individual, each designed to support their particular failing bodies as the mind became no longer able to command muscles to twitch and stretch or for the senses to locate the body spatially in relation to other objects or even to itself, and each was appropriately cushioned to prevent pressure sores developing.

All food had to be pureed as the swallowing reflex had become incapacitated – dysphasia as it is clinically called - and even drinks had to thickened and spoon fed because of the attendant choking risk. All this was new to me.

I had been prepared for the radical departure from the light support requirements I had been trained and asked to perform so far to the total care I was now experiencing so I was ready for the step change in physical application - but not the mental challenges of seeing individuals so much robbed of their vitality and physical abilities.

The fact that even swallowing was now beyond their control and capacities was a hard fact to face because of the reflexive nature of it, it was like the impossibility of being able to breathe or think – an act so natural that we don't consciously notice it. Neither was I prepared fully for the shocking alteration in individuals from where they once were to where they were now.

The profound nature of this shocking alteration happened upon me accidentally, on my first morning.

Having bathed and shaved a service user together with a more experienced care worker my eye caught a picture on their sideboard – looking out with a crooked playful smile, surrounded by friends was a photograph taken on a group holiday only 2 years before of the service user we were supporting. The attentive eyes, the way they held their body slightly sideways to the camera so they could drape an arm over a confrere and the shy but assured way they tilted their head all spoke of life in the midst of the living of it; for a moment I could not be sure I was looking at the same person as that whom I had just helped to bathe; yet in the set of the face, the proportion of the body – however now slipped out of shape – and those eyes, still gleaming at me now – made it inescapable – it was the same person, but one who had undergone the most dreadful physical transformation. Although my eyes and senses told me this was the same individual, separated in time only by 24 months, the scale of the change was as terrifying as it was tragic. The "Ineluctable modality of the visible" as Joyce put it, "at least that if no more, thought through my eyes" – or, to put it another way - the tyranny of the now - made such change hard to accept.

Yet – as a carer in the dementia field - accept it you must to in order to know and understand the fully rounded nature of any individual you are supporting; without - and this is a critical point that we shall have cause to return to later - *without this total understanding of a full and whole life behind such mentally and physically tortured (and no one who had worked in care could doubt that dementia is a kind of torture of the flesh put through the ringer of the extreme end of mortality) individuals who you now see before you care – real care – cannot long endure.* Without this understanding of who you see before you in their totality of existence, *in whatever mental and physical condition of decimation they now present themselves,* they can swiftly and damagingly become simply a constellation of needs, a concatenation of tasks to be performed upon.

Later when I worked in independent sector care homes this conceptual shift was one of the most difficult to remain fixed upon and something that I have to admit I frequently lost sight of in the high volume and pressurised atmosphere of "commercial care for the elderly" – a term we shall encounter and explore later.

Simply put - from *what was* to *what is* – from history to the present – in dementia sufferers there often lies a huge and almost un-traversable desert for a carer to negotiate if they are to remain focused on providing care that justifies that name.

For all the devastation and veritable horror of the nature of the work, in it could be found the unquantifiable rewards – the caring for someone nearly wholly incapacitated with dignity, respect and absolute care, seeing the interior deep and beyond the devastated exterior, the ability to make contact with something that maintained that thread between the then and the now – all this gave something to you that was more than it took out of you. This is what I went into care in the first place to do, and in it I had found a place where I was reenergised, involved and eager – my bad experiences of care expunged almost totally by the care I felt I was now being asked to give.

The care team there was also profoundly different from what I had experienced up until now, in that they really did operate *as a team*. There was no jockeying for position or pre-eminence, no unseen and unknown resentments and ill will - that appeared to have as little basis as it did sense - and no judgement prior to experience. Further there was no expectation of failure, or the whispering that ensued from it, instead there was guidance, reassurance, informal mentoring and explanation.

That first shift I spent at the Dementia unit ripped past and I deeply enjoyed it and looked forward to my next shift there. When I went in for that shift I encountered the manager of the unit who told me how welcome I had been and how I had impressed my fellow workers – for once in care I felt like I had been validated, as if I was beginning to do things right, and – shock of shocks – to be accepted. Never have I felt prouder except when a shift manager there asked me how long I had worked in care – when I told her it was a matter of months she was taken aback, saying that she thought I had done it for years – that night I cycled home with a smile that even the keen easterly wind could not wipe off.

This is what happens to carers when they find their niche and place in the care field – they become better carers, and, sometimes better people.

Of course I found these odd shifts I was able to take much more enjoyable than my supposedly main shifts. "Service User B" had gone and so while I always enjoyed working with "Service User A" the particular service now felt more and more like a road heading nowhere. Although now the team we had was friendly, young, full of inspiration and themselves inspiring, I was aware that I needed a change.

Tentatively I discussed with the dementia unit manager the possibility of gaining a transfer there. She seemed delighted and said that although there was no full-time position vacant they had long-term sickness issues and holidays that needed covering so that they always needed what was effectively another full time carer to fill the gaps, therefore she said that she would take it up with the area manager at the next meeting.

As a courtesy I thought it was only right that I inform my line manager that I had inquired about a move so that she would not be taken unawares, and asked whether she would support me in this. I thought it was reasonable request and one that would tick most boxes – she would not have to make any staffing reduction in the service I was presently in if I were to voluntarily leave, and I would be able to move to a service I was enjoying working in and appeared to fit into. Therefore when I was met with what amounted to a point-blank refusal I was somewhat taken aback.

There seemed no reason for this denial, so I inquired why this was, as I thought it only just that I be given the grounds of a move being ruled out – especially considering the apparent tenuousness of my position; I was told a move in the immediate future was out of the question because I was needed in the service in which I was employed in - which I found curious and fundamentally untrue as we were clearly now over manned, and as I was asked on numerous occasions to fill in for other staff holidays in other services where I had no knowledge of the care needs of the people I was supporting and felt like I was being "shuffled around the board" simply because we now had too many CW's to support one service user.

However I left the particular issue of an immediate transfer there, despite me having received no satisfactory answer, but I did ask what would happen once "Service User A" was transferred off the site to another location – would it still be possible for me to support him in terms of location? To this I was given the frosty response that I was concerned about something that was none of my business – which was strange as this factor so obviously had a bearing on my future employment – and that any prospective move for "Service User A" would be some time off.

I didn't push the issue here but was hardly satisfied by the answers I got, the only firm request I did make was that if there was any cover needed at the Dementia unit, and if it was possible regarding the support that "Service User A" needed, that I be considered for them; this was waved away with a half-hearted nod.

Even though I felt I had been fobbed off with lame excuses rather than solid reasons for why a transfer was out of the question I tried to understand the position from my line manager's point of view and took the answer that I was still needed in the service I was in, and that my position was secure - at least in the near future - at face value.

To say I was somewhat surprised would be a chronic understatement then when I learnt only a day later that the service provider was not only actively looking at an outside placement for "Service User A" but that they had in fact had narrowed their choices down to 3 locations – all out of the area, all virtually impossible for me to reach by bike – and a final decision was expected within a month or six weeks at the outside as clearance was only awaiting approval at regional level.

All this was known to my line manager when I had had the conversation with her 24 hours before; what I thought had been a fobbing off had in fact been point-blank lies. Worse was to come in the next couple of weeks as I found out that cover for the Dementia unit was being provided by other CW's who hated going there and would - and did - everything to avoid having to do shifts there (including a CW who had a chronic back condition that made any slightly strenuous physical effort painful and aggravating to their condition and another who had just returned to work after a double hip replacement), all the while I was being farmed out to the services these other CW's were being removed from to cover shifts at the Dementia unit – in short I was covering them while they covered the Dementia unit short staffing.

This struck me as perverse to the point of doing an injustice to the word as it seemed to defy all logic in man-management.

Unbeknown to me though I had transgressed the one golden unwritten rule in care, and that is that CW's should never think autonomously, never ever vocalise those autonomous thoughts and most of all must never ever, ever try to put those words into actions.

Another CA told me of the management mindset at the service, I recall them exactly because they were so striking.

"If you ever come up with any ideas of your own they will do anything and everything to not do that thing what you thought of [sic], not even if it's a really good idea. They just don't like you to have ideas."

It became clear to me over the next few weeks that even if every other CW in the service were either struck by ball lightning, hit by a convoy of steamrollers or crushed by a slab of frozen urine dropped out of aircraft toilet then I still wouldn't get a position at the Dementia unit.

The reason? I had broken the above golden rule – I had had an independent thought. Worse, I had tried to act upon it autonomously; I had *initiated* an action rather than been *instructed* to undertake one. This was akin to crossing the Rubicon; rather than just a rule you could break and come back from; it was a fatal error.

This golden rule was and is not restricted to this particular care provider, it is replicated right throughout care and will be familiar to many a CW - and many more former CW's - as it is one of the prime causes of why so many carers leave the profession so disillusioned and despairing.

Imagination, inspiration, autonomous thought and free thinking are enemies of care management, and those that have them are enemies of management itself.

Quite how much free thinking is abhorred became crystallised over the next couple of weeks.

CHAPTER XII

Thinking Is Not Free

The situation I found myself was one that many CW's will identify with, the inability to be able to make career choices and directions and to be supported into them is sector wide, as shown by the observations of both the Kingsmill and Cavendish Reviews quoted in the previous chapter. CW's are expected to simply receive orders and to act on them, they are never supposed to display anything like ambition or even an idea where they would like their working lives to progress to.

But this attitude connects with a deeper theme – that people occupying certain jobs of a perceived skill level – that is those in what are called "low skilled" occupations – are not meant to have imagination, never considered to have a rich and full interior life and most definitely are not supposed to have an eagerness to improve either themselves or their experiences.

In care CW's are meant to be "invisible", silently carrying out care tasks with the minimum of fuss, with little thought and most definitely to never, ever to raise questions or voice opinions about the way that they are being asked to deliver care. In short they are meant to be "servants"

The writer James Kelman summed up this process in his book "Some Recent Attacks" –

"Servants may be heroic or not heroic but they're never fully formed human beings, never particular people."

The denial that anyone should have knowledge, drive and dreams is as corrosive to society as it is to any individual. In care work though this societal manifestation has even more perverse results. In care *any thinking whatsoever* is considered not just outré but *dangerous* to the point of being positively Bolshevik. This was something that dogged my time in care work, blighted many a relation with managers and supervisors, and lead to innumerable travails on my part.

Because I was seen as having something of an education, because I was – heaven forfend! – seen reading in any downtime at work, and that just what I was reading proved of inordinate interest to managers and supervisors the oft repeated mantra was –

"What are doing in care work?"

The idea that having thoughts, interests and an inquiring mind appeared, to anyone in a management position, to be a positive bar to care work. It never occurred that perhaps I was in care work because I actually *wanted* to do it.

Although I was singled out as being especially "suspect" the real picture is that everyone in care has their own dreams about where they want their lives to go and each has these dreams slowly crushed under the wheels of the mindset that CW's are there to be seen but not heard. The low morale that this perpetuates is one of the key factors in the immiseration that many CW's feel and, more pertinently, it slowly grinds out of carers the motive force to deliver good and responsive care.

Nothing illustrates this demotivation than one of my last experiences in supported living for the provider I was employed by.

Having been stymied in my putative move to the dementia unit I turned my attention back to providing the best care I could for "Service User A" who was now alone in the house after "Service User B" had departed.

As the new care team didn't have a car between them "Service User A" was virtually housebound. Because of previous challenging behaviours there were - understandably – strict rules over how and in what manner "Service User A" was to be taken out beyond their immediate environment, however these rules meant that "Service User A" had their liberty restricted to an almost unacceptable level.

It was spring now and the weather was almost as impossibly good as it had been unbelievably bad during the winter; the mornings broke early with a clarity in the sky day after day only ever interrupted by occasional ribbons of cloud not much more significant than fading contrails from high flying passing airplanes.

Coming on shift after a short holiday I reviewed "Service User A's" recent activities; it was depressing for any time of year but especially so with the weather almost demanding to be outside and lived in. Day after day followed the same miserable routine -

"Spent day watching TV"

That summed up their entire existence for the last 2 weeks, never once had they set foot outside the house.

I viewed this as intolerable so raised the possibility of taking them out in a taxi on a jaunt, stopping somewhere and having something to eat, maybe having a bit of a walk, then get a taxi home.

They appeared eager to try it.

Carefully and slowly I outlined the parameters that would have to be observed if this were to be the case; they must be compliant and calm, they must sit with me in the back of the taxi, they must accept the limited destination that we could go to, they must be willing to return when agreed and that if there were any breaking of these guidelines (I was careful to state that these were not "rules" - they were merely the terms under which I was willing to risk my job in order take him out - and that we must both agree with them) that we would return immediately and that this would rule out any further excursions undertaken in this manner; finally all this would have to be agreed to by my line manager.

I laboured these conditions three times to "Service User A" until I was sure they understood. In the response they showed every sign of both understanding and being aware of the conditions hedging the proposed venture.

In suggesting and being willing to take this step I was painfully aware – as noted above - that I was potentially putting not just my job, not just my whole care career on the line, but also the entire scope of possible livelihoods I could seek in the future – if something drastic, whatever that may be, were to happen, it could easily be deemed a safeguarding issue and could result in very severe sanction, including my being put on the register of people excluded from working with vulnerable persons – job death, care career death, making a living death.

[It may appear that I am being somewhat over dramatic when I state the risks I was taking, but let me assure the reader I am being far from it. The issue of safeguarding and the tightrope that all carers walk in simply doing their jobs will be covered more extensively later, for now I only wish the reader to note being put on the safeguarding register means a whole panoply of jobs become unavailable, the scope of which is terrifying]

Taking into account the huge potential risks if the whole venture did go sideways I thought the potential liberty for "Service User A" outweighed personal considerations.

The point was "Service User A" had never been allowed to go "out" in a taxi up until this point because it was considered too risky taking into account their challenging behaviours. However this was based on their behaviours when they first came to the service; behaviour which had since been much improved over time; therefore the stricture appeared to me to have been outgrown by "Service User A's" development.

For a short time I considered just going ahead with what I planned without informing my line manager as I – rightly as it turned out – anticipated that such a suggested course of action would meet with a brick wall of refusal. I have often wondered since how things may have turned out if I had just gone ahead and acted on my own; what possibilities may have been opened, what disasters could have occurred. Either way, caution got the better of me and I phoned my manager and outlined what I had suggested.

From the outset of the conversation I knew the idea was dead in the water, the objections piled up like the driven winter snow I had cycled through in the long dark months of winter -

What would happen if "Service User A" refused to get in the back of the taxi with me?

We wouldn't go, I replied.

What if they started "kicking off" during the trip?

Then I would tell the taxi driver to stop, we would get out and wait until "Service User A" calmed down – the same procedure that was in place if they "kicked off" in a support workers' car.

What would happen if they refused to return?

Then we would wait, until dark if necessary, until they were compliant.

Who was going to pay for the taxi?

I pointed out that "Service User A" had money rolling in with no useful way to spend it, and, it being the fact that the money was to go toward their care provision and support then what use was it if it wasn't to be employed as such in a meaningful way.

I was told abruptly and pointedly that it wasn't up to me to decide just what for and how this money was to be used and that I should mind my own business.

In exasperation I said I was willing to pay for the taxi myself.

There was silence for a moment followed by a blunt single syllable close to the conversation –

"No."

I came off the phone and told "Service User A" the whole venture was a no-goer, that we were stymied at the outset.

I was expecting some type of instant reaction, but there was none, instead they just slunk off to their bedroom. Later CD's rained down from their window – a typical behavioural reaction to feeling depressed. I dutifully picked up the hoisted CD's and couldn't have blamed them if even their whole stereo had exited the window.

It was the first time I had experienced such total frustration in my impotence as a CW in the face of entrenched management or service strictures that appeared to take no account of the individual they were meant to be supporting. It was also my first encounter with the "safety first, last and always" approach in care provision.

It also revealed deeper issues in care, most especially in care provision for adult care service users with learning disabilities – often they are not expected to grow or flourish, they are expected to conform to regular lines of stasis that decides in advance what they may become. This is especially true of older service users like "Service user A" – they are channelled into "interests" or "activities" that they show no inclination to enjoy or partake in, and activities that they potentially could benefit from are barred from them by the corral of tried and tested practices and restrictions often based on a long out of date evidential basis.

This last factor leads ossified care planning and of a philosophy that places safety before all other considerations – if a certain activity is deemed "too risky" when an individual arrives in care often this restriction remains in place long after the reasons for it no longer pertain.

The irony is that the notion of "Person Centred Planning" (known under the unfortunate abbreviation PCP – oh that it would have the restless dynamism associated with the drug of the same term) is preached constantly in care, it is thundered upon in training, emphasised in service providers marketing buff, and is a key requirement of the CQC; and yet it is honoured more in the breach than in the observance. Time after time I have witnessed service users in many different services display an aptitude and interest in something that remain untapped simply because it is not in the care plan.

More importantly I have seen the needs of service users go unmet as they change – this is especially true in dementia care. Often a decline in capacity and an increase in needs of service users run far ahead of what is care planned leading frequently to unnecessary pain, distress and sometimes real and prolonged suffering.

Care plans are meant to be – in theory - responsive to changing needs and development, instead - and in practice – they form restrictions or methods of care that neither meet personal development or respond to increased physical and mental demands. Rules and practices that were once applicable and indeed eminently sensible become set like concrete in care plans; in other words they become versions of the fable of The Guru's Cat.

[For those unfamiliar with this fable it goes something like as follows - When the guru sat down to worship each evening, the ashram cat would get in the way and distract the worshipers. So he ordered that the cat be tied during evening worship.

After the guru died the cat continued to be tied during evening worship. And when the cat died, another cat was brought to the ashram so that it could be duly tied during evening worship.

Centuries later learned treatises were written by the guru's disciples on the religious and liturgical significance of tying up a cat while worship is performed.]

This final frustration proved too much for me, I felt I could not provide the care needed for "Service User A" and that any fresh ideas I may come up with would be stifled.

Besides I wanted more and more desperately to work in dementia care, so I made the decision to leave.

I applied and was interviewed for a job in a local nursing home and was offered a job, so I tendered by resignation, served my notice and entered the mad realm of for-profit care for the elderly – perhaps the biggest blunder, and yet the most revealing and bleakly enlightening move of my largely misbegotten work career.

PART II

Descent To Hell

"But now my more fruitful time has turned its back on me, and old age comes, with tottering step, that must be long endured.........The time will come when the passage of days will render such body as I have tiny, and my limbs, consumed with age, will reduce to the slightest of burdens. I will be thought never to have loved, and never to have delighted....... I will go as far as having to suffer transformation, and I will be viewed as non-existent, but still known as a voice: the fates will bequeath me a voice."

Ovid - Metamorphoses.

CHAPTER XIII

Free For All. Costly For Everyone

If adult care is the wild west of social care in the UK then care for the elderly is its Deadwood. No part of the social care field is as dysfunctional, poorly organised, inadequately regulated and subject to the most appalling working conditions as elderly care. What should be a place where those whose care needs can no longer be met in the community and their own homes and who then are so compelled through circumstance of afflictions to pass, what is in effect, their last years or months in a comfortable, loving and compassionately guided atmosphere are instead subject to what could be described as battery farming for the slowly declining. The fact that care for the elderly is the most lucrative opportunity for for-profit care companies to exploit means that elderly in care home are in effect regarded as ATM's with wrinkles. Forced into completely unsuitable living arrangements – often living cheek by jowl with individuals much more unstable, mentally ill, personally intrusive and frequently violent than themselves the elderly, already suffering from, in the main, confusion of their own, find themselves condemned to a bedlam so disorientating that many are reduced to an anxiety filled existence whose only refuge is barely possible sleep.

The care that is supposed to be more intensive because of their increasing needs is instead more inattentive, more distracted and less compassionate than that they have experienced at home. Overstretched and overstressed staff have little time to actually care as most of their time is filled with more urgent and immediately imposing concerns, such as feeding and hydrating those unable to do so themselves, attending to personal care of the incontinent and trying to preserve a modicum of safety for residents from the depredations other residents (physical attacks of residents on each other are commonplace) and from those of their own deteriorating conditions (falls, choking fits, sudden deterioration of their underlying illnesses).

This would be a mammoth task for a well-trained, motivated, engaged and valued workforce, but for one that is mostly skills and experience poor, sometimes inattentive, often demotivated, constantly devalued and, frankly, on all too many occasions, uninterested past the point of neglectful, care homes often resemble something approaching a psychiatric war-zone.

Care for the elderly is, by and large, the sinkhole of carers, the place where poor care is not just tolerated or unremarked upon but often covered up either by other carers or – much more frequently than people would dare to imagine – by mangers. Although outright abuse is, in my experience, rare, and very serious abuse even rarer (I personally never encountered it) neglect is endemic. Almost every day I worked in elderly care I encountered residents left soaking and stewing in their own urine and faeces (several times I found one particular resident so wet that they had soaked through their incontinence pad, through their clothes, through a cushion on their wheelchair, through the fabric of the wheelchair seat and onto the floor) underfed and/or dehydrated, lying in bedding that had faeces or food stained into it, lying on untreated bedsores, the faces of those who needed assistance with feeding half covered with the remains of food and residents "parked" in darkened rooms alone and with no stimulation strapped into wheelchairs.

I often asked myself how people notionally supposed to care could be so lacking in empathy and compassion that they could, seemingly without pangs of conscience, leave other human beings in such a state. But then I had failed to take into account that the lure of Coronation Street, Emmerdale or EastEnders is that much greater for those on minimum wage than that of seeing that their job was done to any degree of adequacy.

While such horrendous neglect is always inexcusable sometimes it needs to be put in context - often care staff are so stretched that they simply cannot address these issues; on almost every shift during the day in nursing homes I worked in manpower was tested past breaking point by one or two absentees, mostly through sickness, leaving what had been barely adequate staffing unable to meet the needs of all the residents they were caring for; when one or two absentees marks a degradation of total staff by between 15% and one third, decent care can hardly be maintained; frequently I have seen staffing on shift fall by almost 50% making not just care impossible but the very safety of residents also.

Yet failing care staff is only half the story, much more weight of responsibility for the poor standard of care in care homes falls on nurses. Few are the nurses that I have worked with who are not only able to do their job but do it with alertness and efficiency; far more common have been nurses who are too lazy, too badly trained, too distracted or too incompetent to do their job. Poor nursing has accounted for most of the neglect in care homes that I personally encountered for two reasons.

A nurses job is to not only see that residents medical needs are attended to - from administering medication to monitoring their general health, to hands on nursing such as dressing of wounds and applying first aid - but it is also to make sure that care staff are doing their job – staff failing in their basic duties of care should be accountable to the nurse and they in turn should be on top of any lapses and failures.

Over and over again though I have taken over a shift where nursing leadership appears to have been either completely missing – allowing care staff to do as little as they wanted – or wilfully blind in the failing of basic care – such as seeing that residents are properly hydrated and fed and that they are not left soiled and that they are not a danger to themselves or others.

While often failing to monitor and counter poor care it is in this second function – the authority of nurses - that perverse results often obtain, for it is well-nigh impossible for CW's to counter or complain of poor nurses as a nurses word is almost always taken above that of a CW. Worse, a complaint against a nurse can lead swiftly to a termination of employment for a CW at that facility – this is something I personally witnessed not just once but several times.

Worse still is the fact that the abuse and outright neglect that I have seen, or that has been reliably witnessed by other care staff with corroboration, has all been perpetrated by nurses. On one occasion a senior nurse effectively assaulted a resident and this was witnessed by another nurse, when the witnessing nurse reported this abuse to management they swiftly *found themselves* on the wrong end of a disciplinary on trumped up charges, and even though these trumped up charges were dismissed following an investigation, within 2 weeks of making the complaint the complainant was forced to resign because of "a breakdown in trust" between them and their colleagues.

In another example a general nursing resident in their 90's was left with unaddressed, undressed and untreated bedsores for weeks on end, despite the impassioned and repeated imploring of a CW caring for them, until eventually they were rushed to hospital with multiple and serious infections, chronic weight loss, dehydration and anaemia, all these issues – bar the dehydration - which still technically was the nurses responsibility as the resident was supposed to be carefully monitored as to their food and fluid intake – were the provenance of nurses who were too idle to do their jobs. The resident was discharged from hospital in a couple of days and died three weeks later. In a bitter twist the carer who had done most to seek medical intervention for this resident was blamed for the condition in which the resident was admitted to a startled and horrified hospital, simply because they had not put their concerns in writing.

The CW in this situation was in a no-win position as to put their concerns in writing would have been to bring the wrath of the nurses responsible down onto their own head; this is because, in care homes, nurses are, to all intents and purposes, a law unto themselves. In the worst examples they have ruled by fear and intimidation to such an extent that no one dare make any claims against them for dread of such reprisals as mentioned in the above.

Even in less intimidating circumstances nurses can still safely ignore the concerns raised by CW's by simply turning a deaf ear to their invocations, eventually wearing CW's down by their inaction.

At one home I worked in a new starter nurse of vast experience was horrified by the uninterested nursing and declared that the CW's were solely responsible for the decent care at the home and of its continued maintenance, as the nurses had totally absented themselves so much from it. They further went on to say that if the few dedicated CW's at the home left then the home itself would be closed in short order by the CQC – a horrendous analysis and a perverted inversion of what is supposed to be occurring in care homes – nurses being applied and involved. This is one of the ultimate ironies that I found in care for the elderly - that a CW led establishment – something that had been blamed for the chronic abuse and failing care at Winterbourne View, and the deleterious effects of which I have already discussed - in some cases was the *only* way that an establishment stayed on its feet.

Managers of homes have, in my experience, been complicit in the covering up of failures in nursing - most frequently, as above, by shifting the blame onto hapless CW's This is partly because, by law, nursing homes have to managed by a registered nurse so they have an interest in protecting their fellow nurses even at the expense of care for residents. Managers also, generally, have a dim view of the CW's that they manage, they treat them with something between indifference and open contempt, flouting many of the conventions that govern most employment hierarchies, such a speaking to subordinates with any modicum of politeness or respect, taking disciplinary measures behind closed doors and keeping such actions and consequences confidential, and regarding CW's as human capital to be conserved every bit as much as the financial or fixed sort.

Dysfunctional management is the most common reason why homes fail, and in ways that are not always obvious. While a manager must be on top of the all the care provided in the home – from the quality and nutritional value of the food, to the safety and wellbeing of the residents – and bears ultimate responsibility when any of these fail – they must also see that CW's are able to do what is a very difficult, mentally and physically draining job in an environment where they are valued and encouraged. The trouble is that that this approach demands nuance, intelligence, empathy, acuteness and vigilance – in other words it takes time and effort; most managers either see little benefit from this or simply do not wish to try it. It is far easier to bully, cow, intimidate and threaten; but in the long term this method is self-defeating – good care staff get driven away and poor care staff stay because they either are so disconnected from the work they do that they do not mind the regular rollockings, or else put up with them because they know they will struggle to find jobs elsewhere. Eventually good care staff are replaced by ever poorer staff as a facility rushes to try to plug holes left by departing CW's and eventually care degenerates. This is a form of slow death for a facility and one of two things happen - either residents leave and potential new ones are put off by the declining reputation of the facility, or the regulator steps in to close it as care failings become critical.

The truly depressing thing in all this is that it need not be this way, simple steps can be taken that can address all these failings at no extra cost. The sourcing and retention of good nurses, the valuing of good care staff and the promotion of decent care by example rather than through fear can all create a virtuous circle – a happy work environment lowers staff turnover, attracts good staff to replace any that move on, and an atmosphere of collective aims and ambitions can raise all boats. Yet the will is totally lacking and this comes from the structure of care – a mainly for-profit one. Results are expected by the quarter, return on capital rules and human costs have no entry on a balance sheet. As for resident care, the pressure of demand will almost always fill holes - in all but the most dysfunctional and failing service - where they occur. With an ageing population with ever fitter bodies yet ever more fragile minds there will always be the aged and infirm looking for residential nursing accommodation – where demand is exponential there is no way of improving the quality of care as only the quantity of "throughput" of paying clients is what is sought; as long as the tills keep ringing owners have no interest in actually seeing that good care is provided. The same applies to CW's – where recruitment is effectively picking people off street corners to be carers it matters little about staff turnover – as long there are sufficient bodies in a building to meet the minimum staffing requirements then it matters little what value those staff bring.

Just how easy it is to become a carer in a nursing home will become apparent in the sequel.

CHAPTER XIV

A Poisoned Tree

"Many experts interviewed by this Review have said that they do not think it is possible to train people to be caring, if they do not start off with the *right attitude and aptitude* [My italics]. To get the right quality of care, it is therefore vital that the right people are recruited to caring roles."

The Cavendish Review.

The problems with care for elderly, most especially in the private sector, start right at recruitment. The lack of rigour, screening or assessment as to suitability of applicants for work in the care for the elderly field – and most especially in dementia care - or oversight of the whole recruitment process individually in different care providers, and across the care for the elderly sector as a whole, is frightening.

I have already remarked at length over the deficiencies that exist as to screening prior to employment in care, yet it is worth reiterating - as specifically in reference to the care home sector - because of certain additional concerns that are raised.

My own experience of job hunting is salutary in this regard.

Once I had made up my mind that it was time to move on from the supported living service that I had been working in, toward one in the dementia care field I, quite literally, did not have to actually look far.

12 months earlier, when I had been clubbing away at applications and scouring the jobs pages of the local press for care jobs, a brand new care home had opened just down the road from where I lived. I had actually applied for a job there back then but received neither a reply or any other response so I had carried on looking.

Now, actively looking for a job again I called in to the home to see if there were any vacancies. The pleasant Administrator handed me an application and then showed me round the facility.

It was impressive, no expense seemed to have been spared, all the residents' rooms had built in televisions and on suite bathrooms, the furnishings were neat and homely, the dining rooms were well equipped and the whole building seemed to hum with quiet efficiency – CW's flitted backwards and forwards with a well marshalled fluency and nurses were busily getting on with the morning medication rounds. All in all it looked like paradise – as if here was a dream opportunity parked right on my front doorstep.

I took away the application and filled it in then spent the next few days desperately hoping that the opportunity would open would open up so close to home, so close to my desired field of care work.

"Everyone gets what they want" goes the line in Apocalypse Now, and to paraphrase - I wanted a job in dementia care, and for my sins they gave me one.

I came home a few days after filling in the application form and found a letter there from the local home offering me an interview. I arranged it as fast as I could and a couple days later I found myself in the foyer of the care home waiting for the manager and Administrator of the home to take me in to interview.

During the time waiting for me to be called in one or two of the residents who slid around the sleek interior of the ground floor of the home like disinherited ghosts drifted up to me.

Their faces became what would become a familiar image over the next several years that I worked in the care for elderly field. Pale skin from lack of direct sunlight, their cheeks drawn into a Munch-like scream elongation by the removal of their false teeth due to dysphasia, their speech a disconnected thicket of words - picked over half remembered grammatical constructions that ran into one another like clumsy supertankers colliding in an empty sea - gestures slow and fastidious at times, at others so shockingly swift it was as if they had been jolted by an unexpected electric impulse – so swift in fact that I admit I reflexively flinched once or twice as wild flapping arms flittered alarming close to my face.

The most shaking impression though was that made by their eyes – and what could be called the "mental slippage" behind them – a constant look - similar to someone who has misplaced something very valuable and the knowledge of which is just occurring to them – the – "Oh my God no, not that; I haven't lost that have I?" moment of stunned shock we all recognise, before the feeling becomes crystallised in the absolute horror of the full knowledge of the lost thing. Only that full horror is never fully crystallised in dementia sufferers, never becomes the full knowledge, only there, ever, is that creeping realisation that "something" precious has been lost, always constantly on the point of occurring and always recurring – this point perhaps the long minute of the present - with no past or future apart from recurrence of what can be held at this point - that Nietzsche famously referred to.

The pain and the only slight amelioration of dementia is the fact that while all may be lost the nature of what has been lost – the full quality of the mind – is only transitory, a moment, a fragment of terror that is suddenly washed back away by the receding tide of forgetfulness.

Edgar Allen Poe was on to something when he described a similar effect of his alcoholic drinking -

"I became insane, with long intervals of horrible sanity."

There was, and still remains for me, in this aspect something of Yeats' terrible beauty existent "when things fall apart" - the indomitability of the human consciousness as it persistently and desperately seeks to cling to the increasingly treacherous ice-wall of perception and connection to the the world around them. It is this "look" that ever reminded me see that all of what a dementia sufferer once was still existed in a fragmented form behind the often inscrutable or unfathomable depths to which they may descend.

I made stifled and lumbering attempts to converse with them and in my own clumsy way felt like there was something inside of me impelling me to reach out to these broken-minded individuals, a want to find something of them behind their obvious limitations in finding themselves.

Is this what I was meant to do I wondered? Was the will to find what they could not find themselves enough to make me a good dementia carer? Could I cope with often not finding a level with which to interact with them?

These questions I would often return to, and interrogate myself with, over the coming years; and, to a very large and unsettling degree, these were the questions never asked of in any training, never required of me in any review of my work, never tackled with those I often came to work alongside.

Most of all they were never once raised in the interview that was soon called into.

This interview – such as it was - couldn't have gone better, though I did wonder how it was possible to make it go badly, all I did was to mention again the buzzwords - treating service users with respect, dignity, kindness, understanding and care. To every question I considered the answer but they were so obvious that at first I thought they were trick ones with some mystery answer only the initiated would be aware of, but no, they really were as straightforward as they appeared.

Later I was told by the administrator, whose background was personnel that my interview was the best she had ever taken.

Jesus, I thought, what did the others go like? Even more, how was it possible to fail such an interview? suggest that the best way to better care was to beat service users until they begged for mercy or stopped breathing? Heavens, all I did was apply the same vocabulary that I had been briefly coached into by my friend when I first took a care job.

The only curveball came near the end – what was I to do if a service user became challenging.

Eh?

Challenging?

These were old folk, septuagenarians, octogenarians, even one centenarian, how challenging could they be? What were they going to do? Batter me with their zimmer frame?

Despite what I thought was the most bizarre question I had ever faced I gave the stock answer of leaving the area where they were being challenging, making sure they, and any other service users, were as safe as possible, and allow the challenging individual time time to calm down on their own.

How amazingly dim I was, how stunningly ignorant, not to say arrogant, I was. If I had had a glass head the manager and administrator would no doubt have laughed themselves blind.

At the end of the interview they asked me to leave the room while they talked privately. Sat in the foyer, once more making hesitant but earnest conversation with the same service users I had tried to converse with before, I inwardly was secretly praying that things would go my way.

After no more than 5 minutes I was called back in, and, totally unexpectedly, offered a job.

I was stunned.

No long deliberation, no discussion up the chain of management, no extra questioning, not even being told there was no position as yet available but they would call me or consider me when one did open up, much less being told they would inform me of their decision over the next couple of days – nothing, in short, to suggest any sort of in-depth discussion about my suitability. 5 minutes was all it took for them to decide that this total stranger - a walk-in off the street less than a week ago - was worthy of working in one of the most challenging sectors of care work.

The main point to draw from this personal experience is just how easy it was – and still is - to get a job in a nursing home.

While, in receiving the offer of a post in the care home, my references were checked, this only occurred *after* I began my employment, in other words they had no knowledge apart from my own word as to any skills I possessed or whether I was actually any good as a carer; yes I was subject to a SOVA (Safeguarding Of Vulnerable Adults – a register of those banned from working with vulnerable people, now folded into the more comprehensive DBS check that replaced the old CRB check) and (paid out of my own pocket) a CRB/DBS check but there was no other "due diligence" check to look at my suitability, no requirement as to proper prior training and experience of service users with dementia, and no proper oversight of what was supposed to be a 3 month probationary period once I started.

Instead it seemed as if all they really wanted to see was if I could parrot out platitudes, nod in all the right places and have two functioning arms and legs; or, as a cynical nurse a few years later in my employment at the home (in my experience nearly all nurses working in social care are cynical to high degree) put it when asked how a particularly lazy and uninterested new CW had got their job, they replied –

"It's simple; they've got a pulse."

Although this comment was particularly sour they weren't too far from the abject truth, as it turned out that my interview was highly rigorous compared with what the recruitment policy in this home degenerated into. By then potential employees were hardly questioned at all about their experience or asked to supply suitable, *or indeed any,* proper references. Many carers who joined the staff after me not only could not provide decent references but in fact *had been fired from previous positions* due to their care failure. Others had been moved on from other homes within the company because of complaints or their lack of aptitude for care work, and still others simply did what I did - drifted in off the street and asked for an application form – some of these individuals with no previous care experience, were allowed to start work only a few days after their interview *while waiting for their DBS clearance to come through.*

Just to re-emphasise that point – care workers were taken on and allowed contact with service users - frequently on a one-to one basis, and with no other carer present to monitor them - *without the home knowing if they had previous convictions that should bar them from working with the vulnerable.*

That this was not only common practice but allowed to go on with the knowledge and assent of the manager should disturb anyone with an interest in elderly care. The fact that this practice contravenes CQC guidelines and prohibitions made no difference, as the home by then was desperate to fill vacancies, and the chances of them getting an unannounced inspection were remote. In other words the reward – getting bodies in to meet staffing ratios - was worth the risk – potential serious harm or abuse of service users.

Just consider that point for a moment, as it deserves some thought and indeed should provoke much soul searching as to what a pass social care has come to. People with no previous experience of working with the elderly and dementia sufferers were allowed intimate contact with the highly vulnerable *without any oversight whatsoever.*

Further, ask yourself would you entrust your children to such an environment? Would you leave your kids with a service that singularly failed to carry out the required basic checks to see that they were not exposing them to an intolerable level of risk?

I suspect that if this were the case then the public outcry would be substantial and calls for immediate action deafening.

And yet here we are, leaving our aged, seriously ill and mentally incapacitated grandparents or parents, brothers, sisters, uncles or aunties to a system that tacitly – by default of poor oversight and service providers only interested in the bottom line of a profit and loss account – condones such practices.

Once more the only conclusion we can draw is that which the Financial Time came to some years ago – that we actually care very little about the the care of our elderly and mentally ill.

A further complication and failure of proper oversight is in the number of oversees CW's who are employed in the care home sector.

What most people should know to begin with is that the care home sector would grind to a halt or become financially unviable without the use and aid of oversees workers. Enter almost any home in the country and you will find a substantial minority - or in places an overwhelming majority - of carers will be non-British born. I have already commented about the United Nations aspects of care and will return later to the deep complications and conflicts among staff that this creates, but here I limit my comments to matters respecting pre-employment checks; the operative and immediate question here being - what use are background checks of past criminal convictions when someone may have come from the other side of the world? What use are SOVA checks if individuals have little track record in care, not because they have never worked in it, but because they have never worked in care *in this country*?

This is not a Trojan horse argument for the prohibition or limitation of overseas workers in themselves – although such an argument could be reasonably made (and will be considered later) – it is simply a worrying phenomenon that the very checks we rely on to keep vulnerable adults safe have limited or no value in regard to overseas workers.

Without resorting to hyperbole, would someone only recently moved to this country be allowed to walk into a job where they dealt with state secrets? Would we allow private companies to actively recruit overseas and to bring in workers from the Far East, the Indian Subcontinent or Africa, specifically for work with sensitive information as to national security issues with no other criteria than they were cheap to employ? Would we allow non-British nationals access to such sensitive information knowing that the limited background checks that exist are of next to no value?

The very suggestion is ridiculous, yet at the same time we allow such immigrant labour to work with some of the most vulnerable in our society.

Such an argument obviously raises sinister undertones of closet racism, or at least and-immigrant dog-whistling.

This is to confuse rather than address the salient issue.

I have worked with just about every nationality you could mention and found that whatever someone's country of origin it made no difference as to their competence or their quality of care – good carers are good carers, bad carers are bad carers and the colour of the passport or skin is no surrogate as to their caring capacities. However we cannot let the valid and vital point at issue here - the limited value of background checks for those overseas CW's - to be obscured for fear of possible offence.

It can be argued that recruitment failures in care for the elderly is the poisoned tree from which all manner of ills result. The poor standards of recruitment means poor carers are often employed, poor carers mean poor care and poor care leads to real suffering, discomfort or at least inconvenience for people right at the point when they most need an attentive, motivated and applied workforce.

In the subsequent I will go on to recount many incidents of the above and nearly all of them come from the point of lax standards in screening carers.

The fact that this is remains the case after countless episodes of well publicised, endlessly reviewed, reported and lamented over failures in care is something that should shame us all.

Perhaps the fact that it doesn't is the most worrying of the conclusions to be drawn from from this observation.

CHAPTER XV

Challenging Times

Nothing can prepare any care worker, no matter how experienced, no matter how long they have been employed in care, for their first contact with private sector care for the elderly.

It is like voyaging down into Dante's Hell - the intensity, the sheer volume of demands that are placed on staff and the claustrophobic atmosphere all contribute to a very difficult working environment; the fact that often care homes are run right down to the wire of adequate staffing only adds to the pressures that build over the typical 12 hour shift.

The stresses of working in such a atmosphere places huge strains on staff that nearly always go unrecognised until they become clearly and painfully apparent.

It is not surprising to me that nearly all the incidents of poor or abusive care that are caught by families or regulators or the media occur in residential care for the elderly. This is not to excuse poor care or abuse – which is always inexcusable - but without taking into account the significant difficulties in the residential care sector for the elderly facing staff - who are often overworked, undermanned and sometimes poorly trained or prepared for what is a highly specialised labour - we are condemned to repeatedly see such abuses reoccur.

A proper evaluation or empowered review of how we care for our elderly and how that care is delivered is not only overdue but imperative. The current system is at breaking point and now may be our last best chance of reforming care for the elderly while we still have time to do it in an orderly fashion, rather than when we hit the wall that is looming not so far down the tracks.

To demonstrate why such a review is needed we need to look at the various degrees of care that the catch all term of care in nursing homes covers. Most homes will typically offer three divisions of care – Residential, Elderly Mentally Ill (EMI), and Nursing.

Residential Care broadly means those who can live independently but who can no longer live comfortably or safely in their own homes. The classification includes those who have physical disabilities that render their needs more intensive than can be met domestically so need 24 hour nursing assistance on hand to meet their everyday requirements. Typically they will still possess decent mobility, be still able to feed themselves, have capacity to make decisions about their own welfare and be mentally stable and engaged in their surroundings but they may no longer be able to manage all aspects of personal care safely, may require prompting or assistance with the taking of medications and need to have cooked meals prepared for them to meet their required nutritional intake.

Also classed as residential are those who are in the early stages of dementia and who therefore cannot safely be cared for in community – they may be confused, prone to wandering, incapable cognising the immediate (they may think they still have children, have a job, have a departed husband still living and prone to fabulation – creating defective truths anchored only partially to reality) and unaware of their immediate situation and surroundings, however they will still be able to, say, attend to basic personal care – i.e. wash themselves, clothe themselves, toilet themselves - but may need prompting or some minimal assistance for all these tasks.

Finally there are those with some combination of both these care needs – they may need assistance for mobilisation and suffer from some early stage dementia issues.

The problems of caring for Residential Care service users should be obvious, the very mix of those with only physical disabilities with those with only early stage dementia or a combination of the two can make for a volatile environment and demand carefully calibrated care responses. As a simple example a service user with little or no mobility may well object forcefully to an individual with early stage dementia wandering into their room but is in no position to do anything about it themselves so will need care staff to maintain their privacy; yet without isolating them from all other residents. Equally an early stage dementia sufferer may be perplexed why they cannot simply wander where they wish for they may believe they are still in their own home and resent not have complete liberty of movement; and someone with combination of early stage dementia and physical disability may need balanced care to address both needs equally.

CW's working on residential units may find more of their time taken with conflict resolution than actual care, they may also have issues around the flexibility needed to address everyone's care needs, from transferring individuals with physical challenges several times a day from different modes of seating/lying and locations (e.g from bed to wheelchair, from wheelchair to comfortable chair; from comfortable chair to wheelchair to dining room chair; from dining room chair to wheel chair to comfortable chair, etc....each task potentially lasting 10 - 15 minutes if hoisting is required) to maintaining such individuals personal hygiene *and* all the while observing, managing and caring for several early stage dementia service users.

This all adds up to a very stressful environment where competing demands mean that no one's needs are fully met. It is not helped by the fact that Residential Care units are usually understaffed as, on paper, many, or perhaps a majority, of service users on such units are able to function pretty much independently, so managers will consequently judge the work as less burdensome, taking no account of the issues exampled above.

Nursing Care is perhaps the one subset in care homes where carers find themselves in the sweet-spot; pure nursing care is the one category where the skills that carers need are fully utilised and employed - caring for people with extensive personal needs are what carers are typically trained for. Although many of the service users may have severe needs – perhaps being unable to feed themselves, hydrate themselves, be doubly incontinent, need all personal care carried out for them, be very ill, have medical needs of varying degrees of complexity - the more simple ones to be carried out by carers – this is what carers not only are best fitted to. It is often the kind of work that they went into

care to do. As a result nursing care posts are those most sought after –
the care may be demanding but it is demanding in a way that carers
can best meet.

If private care for the elderly is hell then EMI nursing is the lowest
circle of it; no work asks as much as carers and as far beyond the limits
of their training and competence as caring for those with the most
acute mental issues do.

Going onto an EMI unit is like stepping back into bedlam. On
my first shift there I was shocked by the noise – shouting, screaming,
howling – and the virtual anarchy of it - residents in a state of undress,
service users wandering freely into other's living spaces - sometimes
when a bedfast service users was in their room - soiled service users,
service users wearing dirty clothes, service users with food plastered
on their faces from the last meal – truly shocking.

Amidst all this were one or two service users who simply looked
shell-shocked by their environment, who were obviously in distress
about their surroundings and whose behaviour was triggered by the
mayhem around them. I wondered how anyone with a shed of sanity
in them could survive in such a setting – in fact what I witnessed over
time was the fact that new service users slid downhill rapidly *because*
of their environment – like some inverted Peter Principal, service
users fell to the level of the other service users surrounding them.
This raises the vital - and unforgivably seldom asked - question of if
mentally ill service users' environment were better would they may
be able to maintain a higher level of mental functioning for longer?
But this question, however important, will be turned to later as it
proceeds from the more presently germane interrogation of what
exactly constitutes an EMI classified service user.

While most of the mental disturbances that I witnessed on any
EMI unit were bound up with dementia this not only does the varying
manifestations of such metal disintegration an injustice through
meaningless reductivity, it also fails address those individuals who
may also, alongside their dementia, have longstanding and chronic
mental health conditions. While EMI may encompass those who are
mentally ill with conditions of the elderly, such as dementia, it also

includes those who have been long-term mentally ill and now happen to be elderly.

This is an important distinction worthy of re-emphasis – elderly classed as mentally ill are becoming more frequently those who have suffered mental ill health for some time *before they became elderly*; therefore CW's are increasingly being asked to perform roles formerly undertaken by trained psychiatric staff.

While most carers in elderly care will have had some (actually pitifully little in my experience – a point we will return to later) dementia awareness training they will have had non in dealing with severe psychiatric cases. CW's are what the title suggests – care workers – they work to assist in the care of service users – *they are not trained psychiatric care staff*, however over the last 10 years their job has become more closely identified with such needs and requirements without training or status keeping up.

The failings here are clear, and obvious – CW's are now being asked to manage conditions and behaviours that trained psychiatric staff may formerly have dealt with, all the while labouring under numbers of staffing that would be intolerable on a psychiatric unit, *and* with none of the support that is available on such units; this simple fact is often the root basis of much of the poor care I witnessed on EMI units. Often I have been asked to manage and support service users for which I simply did not have the skills. EMI nursing often includes people who have severe challenging behaviour issues, among which is the capacity to be very, very violent. After a couple of shifts on the EMI unit I could reflect ruefully on my interview where I was internally amused by the question of how I was to handle a challenging resident –

These were old folk, septuagenarians, octogenarians, even one centenarian, how challenging could they be?

The answer was very

I would have laughed at my former dimness, ignorance and arrogance if the reality was no so grim.

In my time working on EMI units I have been clobbered by objects, struck hard by fists, kicked back and blue, bitten past the point of having the skin broken, spat on, had faeces thrown at me, and, in one memorable incident, nearly been castrated – no joke – after that incident I spent 10 minutes in a cold shower hoping my testicles had survived intact.

There are serious issues here though – how to manage such challenging incidents? Not trained in Control and Restraint (C&R) procedures - because most care companies deem such skills as (rightly) inappropriate and as potentially being seen as abusive (a purely liability issue) - the response most care companies have – such as the one I went to work for - is to walk away from challenging individuals after seeing that they are in as safe an environment as possible both for themselves and others. While this de-escalation tactic is mostly (but not always) adequate in managing challenging behaviour in dementia sufferers, there are some behaviours that cannot be addressed simply by walking away, examples of these behaviours and the inadequacy of the single tool we possess – walking away – in meeting those challenges follow. Walking away is also not only of no use, but not even an option, with the psychotic episodes of some service users, yet as CW's we are expected to deal with these as well.

One service user – a nonagenarian with a long history of severe mental illness - habitually used bowel movements and their consequent issue – faeces – as a way of attracting the attention of as many carers as possible – they would defecate on the floor or in their clothes, or sometimes even in a cup or plate and smear faecal matter over themselves and around their room, then be extremely aggressive during any intervention to get them and their room clean. What would be the correct response here? Walk away? Leave them and their room covered in faecal matter? Leave faeces on a plate? This is obviously an insufficient response, yet I was never given any training to manage this condition other than to be close on hand in case of any bowel movement. With 20 or so other individuals, all suffering from mental conditions of one sort or another, this was clearly impossible. The result was my fellow CW's and I had to deal with the consequences of their mental illness without any of the tools to ameliorate it or manage it.

Another service user I supported was frequently very violent, assaulting other service users or care staff by lashing out wildly and forcefully with their walking stick. While this may sound amusing, the consequences of it were anything but. In elderly care you are dealing with inherently fragile people; while a hefty clout from a stick may be momentarily painful for a fit and healthy individual, it can have catastrophic consequences for individuals who are unsteady on their feet or who have poor skin integrity – liable to tear at the least pulling or pressure - a heavy blow from a solid object like a stick can open up horrific wounds in those with poor skin quality and can easily floor someone who's mobility and balance is poor.

As a CW your first duty is to ensure the safety and welfare of *all* those you care for, the issue of this service user's challenging behaviour therefore raised many dilemmas that are in fact common in the care and management of very challenging individuals.

For those not in care the solution may seem simple – remove the walking stick from the violent individual.

QED right?

Wrong.

No care worker CA or nurse or whoever can remove a walking aid from a service user, for the simple reason that it is, after all, a walking aid – if the item were removed and they were to subsequently to have a fall then serious disciplinary and liability issues would be triggered – I have seen CW's get fired for just this situation.

Another answer that may leap to mind is to keep other service users away from the violent individual; however if the perpetrator is in a communal area this will mean curtailing the freedom of movement and choice of other service users – what is called a Deprivation of Liberty (DOL - a weighty and contentious subject we will return to shortly) – in short you are denying the freedom of non-challenging individuals in order to accommodate an individual who is challenging – a situation which is manifestly unfair and skirts the boundary of acceptability in safeguarding the freedom of service users.

Another answer the uninitiated may spring to is to remove the challenging individual to another place away from other service users.

This actually raises more questions than answers. Firstly what do you do if the challenging individual refuses to move? Secondly, even if you manage to remove them to another place away from other service users you are once more into Deprivation of Liberty issues – the isolating of an individual negating their freedom of movement and choice. Thirdly – if they are mobile there is no way to prevent them from moving themselves back into contact with other service users.

To some the notion of action and consequence should prevail – if someone behaves in an anti-social manner they should bear the consequences of being temporarily removed from the social sphere – this though takes no account of the fact that CW's working in dementia care have to always be aware that they dealing with individuals who are mentally ill – they may not know exactly what they are doing – the challenging individual may feel threatened themselves in some way and their actions completely rational as self-defence – and they may not be able to understand the concept of action and consequence. Normal rules do not apply; what is clear though is that walking away is no solution.

Once more my fellow CA's and I had no authorised or worked-out clear strategy of dealing with this individual, we had no way of either preventing or stopping such challenging incidents. The answer that I mostly arrived at was to place myself between the challenging individual and any other service users – this entailed me on numerous occasions getting battered with the individual's stick for several minutes before they calmed down, but at least it prevented anyone else being more seriously hurt.

Two things should be clear from above – firstly more and more CW's are being asked to act and react to situations that they have no experience or expertise of how to manage; with no clear process or defined strategy other than to walk away they rapidly find themselves enmeshed in spiralling incidents with no viable exit. It is in such situations that abusive behaviour by staff on residents often results – inexperienced, stressed and under huge pressures CW's often choose poor options or indeed lash out themselves. I have been in such situations and it is very difficult to try to resolve them, most especially when you are getting assaulted yourself by someone you are trying to care for. While abuse can never be condoned the poor or absent training for conflict resolution, for psychiatric management or clearly defined strategies to cope with habitually challenging individuals are a huge contributory factor in many episodes that end up in the public eye.

The second point is arguably more direct to the issue of the challenges faced by CW's in care for the elderly, and of EMI care in particular - and that is the dilemmas that CW's face almost every minute of every shift at work. CA's are constantly asked to make snap decisions where right no answer is the right answer and where the best becomes the enemy of the good. CW's are expected to provide excellent care in all situations at all times yet this is simply not possible with the typical staffing ratios that are found in private care for the elderly. This inevitably leads to trade-offs and compromises reached on a utilitarian basis in which all service users may not receive the best of care according to an ideal, *but all receive the best that is possible at the time and in the circumstances prevailing,* to do otherwise is to make care a zero sum game where to give one individual the best of care means that others receive little or none. Another example will illuminate this very issue and also display the paucity of strategies in dealing with challenging behaviour even when such strategies are asked for and received from specialised professionals.

One service user on one of the EMI units I worked in was a habitual night wanderer who would have little sleep and walk the corridors at night – which in itself was not an issue, what was was that they habitually went into other's rooms and disturbed other residents. Even more serious they also had a habit of trying to climb into bed with other residents. This not only goes to the heart of privacy – the integrity of an individualised personal space that is the individual's alone – it also raises the more immediate point preserving the safety of all individuals on multiple levels – including the potentially sexually exploitative (although the behaviour in this instance was never sexualised) - and betrays the feeling of security that all residents in a care home – and their families - should feel about all aspects of their personal wellbeing.

Taking into account all these circumstance and factors it meant that night staff were expected to know where this wandering service user was at all times. However, and this is an issue that is habitually raised in care homes, the night shift operated on a skeleton basis – only half the number of staff were on duty during the night as were on during the day – so it was simply not possible to know where this wandering service user was every minute of the night, as to do so would have meant neglecting all other service users. As this resident's dementia progressed the wandering became more troublesome and the safety issues more pressing. The night staff begged repeatedly that this issue be addressed and so the wandering service user was referred to the Challenging Behaviour Team (CBT) – an outside professional team of dementia psychological specialists. After an assessment all CW's involved with the care of this service user – including me – were instructed to attend a session given by them.

The meeting discussed issues around how this service users' dementia was progressing and why they were acting in the way they did. When it came to practical suggestions as to how manage the constant wandering where it impacted on the intrusion into others rooms the conclusions were jaw dropping. Although an addition of night medication was added to their regular medication it could only be administered under certain strict criteria, so much so that it rendered the medication option negligible, so more weight was piled on behavioural management. The CBT recommendation – the recommendation of the "experts" - was to allow this resident to wander and to prevent them going into other rooms everyone should be locked in their own rooms. This was one of those situations where you are not sure that you have heard what was said properly – did they actually suggest *locking all the other service users up in their own rooms* so that this individual could wander freely?

There was silence in the meeting for a moment as people collectively picked their slack jaws up from the floor where they had fallen before someone pointed out that this violated all safeguards surrounding Deprivation Of Liberty – that in allowing one person to have absolute Liberty everyone else was to be deprived of it in the most violating way – locking them up on a night like prisoners. At this reply it was the turn of the CBT team to fall silent - this suggestion of locking elderly service users in their rooms was their one active suggestion, and that was it.

The night staff were shocked more than most as it was they who would have to deal with the continued and escalating behavioural issues. The medication that had been prescribed did virtually nothing to allow the wandering service user to sleep and because of the strict conditions of its use it was rarely possible to be deployed in any case, so they were back to square one.

Once again night staff repeatedly reported back that this service users behaviour was spiralling to a point where other service users were in real danger, but management were inactive to these concerns; unsurprising as the experts who they had brought in had proved so ineffective.

The situation was eventually resolved, and only fortunately so that it wasn't resolved tragically; one night while the staff were diverted by another service users' needs the wandering service user entered another seriously ill service users bedroom and pushed a pillow in their face, luckily an alert member of staff thought to, immediately after finishing the care of the other service user, locate the wanderer, and caught them in the act. Following this incident all the whistles and bells went off and within 12 hours the wandering service user was removed to a secure psychiatric unit that specialised in managing such behaviours.

There are many lessons to be drawn from this incident, all pertinent and all worth expanding upon.

The first is the almost complete lack of active support by management, clinicians and healthcare specialists that the care staff were given in managing the care of this service user. Without these they were thrown back on their own – limited – resources; without proper effective guidance, without an agreed, broadly implemented proactive and responsive action plan and almost totally without engaged and dynamic oversight. Without any of these the most they could do was to monitor this residents movements and hope that when they were distracted that nothing untoward happened.

The second lesson – which follows on from, and is intimately linked to, the first - is the almost total incuriosity of management in regard to the rapidly deteriorating and night-time situation with this service user. They showed a level of insouciance bordering on the wilfully blind.

This level of uninterest is not uncommon in management attitudes to night staff. Most managers treat night staff even more dismissively than other care staff – a feat of disrespect that almost has to be admired - they assume that all night staff actually do is turn up, throw residents into bed then sit in the lounge watching TV for the rest of the shift. Having worked for a majority of my time on night shift I would be lying if I did not say that many night staff did exactly this, however many more were, and are, excellent carers – like the staff member who's alertness prevented the wandering service user from potentially critically or fatally harming another service user – and who have averted many more potential disasters just like this instance.

Nightshift may be, by the definition that most service users are in bed or asleep a majority of the time, less intensive than day shift, but it has to be remembered that all nightshifts operate with only a skeleton staff – typically less than half that are on duty during the day, sometimes less (I worked several shifts where I was one of only 2 carers spread over 3 floors looking after 40 residents). It also involves periods of high activity - such as early evening, and early morning, when service users are being put to bed or are getting up – where nightshift CW's have to work hard and work independently, knowing that they cannot call on additional help as it is simply not there.

Therefore night staff work difficult hours, often for no extra pay and have to make do with less manpower – the fact that they are mostly ignored and/or chronically disrespected by management and other care staff means that they are never regarded as part of a team – something that critically damages the ethos of "24 hour care" (the "myth" of 24 hour care will be returned to later) that is supposed to guide all residential care establishments – this produces low morale and self-esteem among night workers that even with the best of wills is hardly conducive to a cohesive and responsive care environment.

A third factor is one of the most dirty secrets in adult care as a whole – but particularly concerning in care for the elderly mentally ill – and that is that NHS psychiatric institutions can - and do - more to "chemically cosh" challenging individuals with powerful sedatives than would EVER be countenanced in a care home or any adult care establishment, nor, critically, be acceptable to the public *if only they knew about it.* A short example will be illustrative and depressingly informative not only about this dirty secret but of the failings of EMI care in general.

One service user I knew with early on-set dementia (they were in their late 60's) became rapidly and unmanageably challenging soon after entering one of the care homes I worked in. This declining behaviour had much to do with the rapidly encroaching furtherance of their dementia but equally was not helped by the fact that in their care file it was clearly stated that they hated - and reacted badly - to noise – so they were placed on an EMI unit – a place guaranteed to be filled with a cacophony of the most disturbing noises – work that one out if you can.

Most of the challenging behaviour that I witnessed from this service user was a direct result of external stimulus in the form of noise.

Whatever the reasons they went from undirected fits of anger to attacking CW's and finally attacking another service user – a final straw. Immediately an emergency application was made to have the service user sectioned, a crisis team visited and a section granted. The service user was taken to a local secure hospital accompanied by a nurse and a CW. The CW recounted their experience on returning – at the psychiatric hospital the service user was taken through two locked doors along a dimly lit corridor down a flight of stairs to a basement floor and placed in a room with only a tiny window near the top of the wall a mattress with a sheet and duvet on and that was all. After exiting the room the door was locked on the service user even though at the time they were showing no signs of challenging behaviour.

After a week in the hospital a manager visited to find the service user doing an excellent impression of a zombie, or, in their words - "wandering dead eyed and distant" through the administration of colossal and heavy sedation. On inquiry it appeared that after *one single challenging episode* that they were proscribed such sedatives.

This single example – and I could go on to quote several more very similar ones – raises a myriad of questions – for example - how is it that in a residential setting such medication would never be tolerated yet in an NHS facility – overseen by the same watchdog – the CQC – and supposedly operating under the same conditions, guidelines and philosophy as the private sector - that two radically different responses were applied? Further, how suitable is it that the elderly mentally ill are treated the same as someone much younger and fitter where perhaps heavy sedation might be warranted? And additionally how is it possible to lock someone in a room in the NHS and yet for it to be totally unacceptable in a care home?

If the answer to these questions is that psychiatric hospitals deal with those with extreme mental issues whereas care homes do not, how is it service users with severe and longstanding mental health issues are deemed acceptable to be cared for in a care home and not in a hospital that specialises in these cases? Why was it then that this service user was eventually returned to the care home but all the medication that was deemed suitable for a psychiatric hospital with trained staff on hand was stopped when they were back in an environment where none were available? Finally why was it that in all the care homes in which I worked that they habitually took service users from psychiatric hospitals when their condition was the same or worse than that which saw them hospitalised or even sectioned in the first place?

The answer to all these is money – sadly it is cheaper for the healthcare system to have psychiatric cases cared for by a CW on minimum wage than in a facility with trained psychiatric staff, and *damn the consequences.*

And care homes are only too glad to accept such difficult cases due to cost pressures -

Make no mistake - and this should be emphasised and re-emphasised as it cannot be repeated often enough - this endpoint – the caring for people beyond the capacities of CW's skills - is the product of the for-profit care system that needs to fill rooms and beds to keep the tills ringing.

Most care homes only become economic if they are running at, or close to, full capacity; this means that rooms must be filled at all costs – and all the costs fall on inadequately trained CW's and the vulnerable people they cannot properly care for; or rather homes accept residents without any regard paid to if their care staff can meet the residents' needs or manage their conditions with their current skill levels.

This is problem right across the board in the for-profit care home sector. Over and over in care homes I worked in I have seen residents with extreme needs and conditions being accepted solely on a financial basis with no due regard being given to how they can be adequately cared for.

When I state above *"damn the consequences"* I should be clear just what they are – when care homes accept residents that test CW's beyond their limits of competence, training or caring capacities then vulnerable elderly people are put at real and very high risk of suffering dreadfully, harm *and even death.*

Such dreadful suffering *and deaths* (yes that is plural) was the object in the West Sussex Adult Safeguarding Board (WSASB) Serious Case Review (SCR) into the chronic failings at the Orchid View Care Home. It noted –

"At Orchid View.......some residents were admitted from hospital who were inadequately assessed by.....staff prior to their acceptance and admission. Accepting people who are at the margin of the home's competence and capacity will have a detrimental impact on existing residents as well as the person being assessed............As [Nursing homes] are becoming increasingly important as care providers for people with significant healthcare and nursing care needs, so it is critically important that they have levels of competence to enable them to deliver care."

Let's be clear here – the WSASB SCR makes plain that due to financial pressures Orchid View accepted *any and all residents* with out due care and attention as to how they could be cared for safely so that money could be kept rolling in - *and people died as a direct result.*
This led to the recommendation in the SCR -

"that the process, timeliness and quality of the pre-admission assessment from hospital settings is explicitly tested........with an emphasis on the staffing levels and skills within the home to deliver safe and good quality care."

It notes of this recommendation that –

"Given the pressure across the whole health and social care system this will become increasingly important."

Yet this practice persists, in fact it thriving, it is still common that for-profit homes accept challenging residents on a purely financial basis and this places staff beyond their capacities to deliver even adequate care - *and this will only get worse,* not better, whatever the WSASB SCR may have recommended.
Due to recent austerity measures in the UK the costs covered by Local Authorities (LA's) for social care have been driven remorselessly

down, meaning that care homes have to concentrate ever more on "high value" potential residents – that is residents who come with a large financial assistance package attached to them – in order to balance the books; and such potential residents come with such high levels of funding *because they are so difficult to manage.*

I have personally witnessed this process. Over last 5 years I saw, in the care homes I worked in, the level of demands placed on CW's increase exponentially; whereas 5 years ago care homes may have had 1 or 2% of residents that were very challenging by the time I left care it was up to 30 or 40% that CW's simply could not care for or even manage.

The stress that this laid on the staff was unbearable with the result that staff turnover spiked and, anecdotally, 5 carers I worked with – good, experienced and well qualified carers – left the care sector totally.

This fact – that good, experienced and well qualified carers are leaving the profession in large numbers - acts as a ratchet effect on the prevailing conditions outlined above, as they are largely replaced by inexperienced, unqualified and therefore less good carers, the quality of care diminishes further.

In short we are experiencing a severe and critical skills diminution in the care workforce right at the time when ever more skills and experience should be placed at a premium. This is a recipe for many more Orchid Views, many more SCR's and, most lamentably, much more suffering of the most vulnerable adults in our society.

The final lesson to be drawn on challenging behaviour and its treatment is the other end of the medication spectrum - if the NHS seems to operate on different – more extreme – rules regarding medication then the other end of the spectrum needs questioning – that is, from my experience, certain residents in care homes are under-medicated or, more pertinently, have no plan to deal with their challenging behaviour *other than through medication.*

This is a difficult, tricky and nuanced issue and deserves more expansion, which is why we are heading there next.

CHAPTER XVI

Medication

In one of his most memorable documentaries from during his time working in independent television John Pilger focussed on the plight of mentally ill and how they were treated in the National Health Service back in the 1970's. Seeing footage now is not only heart-breaking but is almost incomprehensible that within the lifetimes of a majority of Britons that those with disabilities were treated in such a callous, Dickensian and incompassionate fashion.

Housed in conditions that now would be frowned upon for a dog shelter, glassy-eyed young men and women sat around all day on a variety of unsuitable chairs stupefied by all manner of brutal drug regimes that also included, besides crude anti-psychotics, psychotropics and opiates that today occupy only places on a narcotics dealers' smorgasbord; all this when the underlying conditions that they suffered from - unstudied by an uninterested medical establishment and so undiagnosed – went unsuitably untreated.

A doctor – stupefied by her own impotence - could barely conceal her shame and horror at the regime she was forced to preside over. Plaintively she related how vicious the over-medication was but that it was the only way in which so many mentally ill patients could be managed in her crumbling establishment, lacking as it did any external stimulus for patients or any active non pharmaceutical treatment. Among many searing images the one of her almost begging John Pilger to bring this situation to public attention, while not saying so explicitly, lingers long after the credits.

We have come a long way in the care of the mentally ill, and so too have we in the care of those suffering from dementia and yet we have also moved surprisingly little in how we respond and react to conditions that still confound a medical intervention to abolish or reverse those very conditions.

Cancer can be arrested, put into remission or even banished from an organ or body, cardiac ailments can be broadly resolved where they are acute, joints and be rebuilt or replaced, yet the mind is still, to large degree, an undiscovered country from whose Bourne few know even how to contemplate navigating into or out of with any useful knowledge. As a result we *still* fall back on ameliorative or remedial drug therapies that in their own way are little advanced from the world that John Pilger found himself shocked by; they may be less brutal and less carelessly administered but still, to a very large extent, they are used to keep those in mental turmoil quiescent, this is still sadly even more true in the case of dementia "treatment".

However much we have failed to change though, we have changed to some degree as my own experience and memory makes clear.

For those of us of an age – anything over 35 will do – many of us may well have had the unfortunate experience of remembering images from nursing homes from 20 to 30 years ago equally as disturbing in their own way as those sown in the Pilger documentary. For myself I well remember being taken by my mother during the long school summer holidays to see a much older friend of hers who was housed in one of the numerous local, local-authority run, nursing homes.

Around this very matronly lady who was alert, quick witted and kind (she kept pressing sweets into my hand when my mother was not looking, or was pretending not to look – as if instinctively aware of the fact that for an 8 year old boy this wasn't the kind of fun holiday afternoon I had been thinking of in term time – and I thought I was masking my ennui *so* well) other residents lolled in their chairs some with tongues hanging out, still others with thin ropes of drool seeping from their lips and yet others – vastly more frightening so therefore striking to me – sat rigidly upright in their chairs, eyes open, staring straight ahead with an unblinking yet unseeing gaze. Although I'm sure this thought didn't register at the time, over the distance of memory this lady appeared to be - among all the other blitzed out residents - like the last survivor of some horrendous apocalyptic old-testament curse, still standing after all others had been taken out.

This particular lady – and I particularly remember her telling me this as she articulated it with such pride as she gestured vaguely at the other paralysed or paralytic drug-gargoyles of human beings – didn't need "any tablets" because – the pointing to her neatly curled and set hair around her temples - she was "all there upstairs".

The tablets that she was referring to soon came on the scene at "tea time", which according to the home routine was 2-30pm; I say "tea time" in parenthesis because precious little tea was consumed. I watched as care workers passed from patient to patient shaking them awake, popping a tablet or capsule in their mouths giving them a gulp of tea and leaving them to loll back while that dose to kicked in before the last had even worn off. No nurse was ever seen and the care workers took the tablets – seemingly at random - from one of several of saucers that were full of multicoloured pills and capsules.

This was dementia care 1970's style – the administration of tranquillisers or opiates for anyone that suffered the least sign of mental derangement or pain – a one size fits all regimen that virtually guaranteed that geriatric environments were some of quietest you could find in healthcare. To say that no one thought this wrong is to be half-truthful but alright – although many thought this blanket medication and over-medication wrong – like the older brother of one of my elder brothers' friends who worked in care and had a tidy side-line peddling tranquillisers not given to already knocked out residents to the town's vaguely strange individuals (vaguely strange was what we as kids called people who we only later we realised were either addicts or habitual drug users) who could see, so he said, no earthly point in giving more tablets to people already comatose (though his opinion was hardly impartial - seeing as he had a vested financial interest lurking beneath his apparent enlightenment) – to a doctor father of school friend who thought the practice barbaric and the use of such drugs as being akin to - and this was the first time and place I heard this term – a "chemical cosh" – but that he was compelled to follow clinical guidelines that authorised the use of such medications as a matter of course, and, sometimes, from the pleading of families that their mentally deteriorating relative be given "something to calm them down".

However most people thought that the routine use of powerful sedatives or tranquillisers as the least bad panacea for a worse illness – senility - as it was called then – dementia as we call it now; they saw nothing wrong with people being reduced to half dead drooling and incontinent shells of persistent existence as being the only way to "control" those who's minds had failed before their bodies and saw this half-life as a natural consequence of age. It also has to be remembered that this was the 1970's – the time when pharmaceuticals were still thought of as so many magic bullets to all the ills of the body or mind, and that the medical profession were never questioned because they were one of the holy trinity that underpinned society – doctor, solicitor, clergyman.

Fast forward to today and we have improved our understanding of mental illness and dementia vastly in a relatively short period of time. Growing numbers of people reaching old age in the developed world has focussed attention on dementia and forced it up the social and medical agenda. medical science and understanding of the brain have revealed some, but by no means more than even the merest beginnings, of the causes of dementia and, let's be honest here, the growing power of the grey vote – older people vote more and are more partisan in their voting as a self interested cohort – have transformed social agendas too. With greater understanding of the neurological causes of dementia has come a greater enlightenment as to how dementia should be treated. The idea that it was best to medicate the bollocks off dementia sufferers to make them "compliant" has rightly been consigned to the dustbin of past embarrassing social mores and medical practices, as has the automatic use of powerful painkillers and opiates unless all other "softer" analgesia's have been tried and found wanting.

Today the hot topic is the use or misuse of antipsychotic drugs. The Alzheimer's Society on their website state –

"90% of people with dementia experience behavioural and psychological symptoms (BPSD), such as restlessness and shouting……..aggression, agitation, loss of inhibitions and psychosis (delusions and hallucinations)…….at some point. These distressing symptoms can often be prevented or managed without medication. However, people with dementia have frequently been prescribed

antipsychotic drugs as a first resort and it has been estimated that around two thirds of these prescriptions are inappropriate."
http://www.alzheimers.org.uk/site/scripts/documents_info.php?documentID=548

The hard number number put on this over – or inappropriate – medication of the elderly with antipsychotic drugs is often put at around a staggering 150 000

To this end the Department of Health (DH) Department of Health has made -

"Reducing the use of antipsychotic drugs for people with dementia….a national priority in England"
(The use of antipsychotic medication for people with dementia: time for action Department of Health. London. 2009a).

Partly as result of this guidance antipsychotic prescriptions for people with dementia have reduced by 52 per cent between 2008 and 2011.
(Audit of antipsychotic prescriptions for people with dementia - NHS Information Centre, 2012)

However The Alzheimer's Society calls for "continuing action" to reduce the total of dementia suffers on antipsychotics. As an alternative to antipsychotic medication as a "first resort" to address the behavioural and psychological symptoms of dementia (BPSD) The Alzheimer's Society suggests that better training for health and social care professionals working with people with dementia.

As American street argot has it – I hear that. Although most care staff are dimly aware that dementia causes challenging behaviour, their awareness of how thoroughly this can happen in causing BPSD is in the main wholly lacking. The most common refrain I have heard from care staff – some of them vastly experienced and well qualified - on observing or being on the wrong end of challenging behaviour is – "they know what they are doing."

This is so common a refrain that I have even uttered it myself, much to my shame and detriment.

Challenging behaviour has been discussed in the previous chapter but it is worth reiterating how psychologically stressful this can be on CW's, never mind the physical battering they receive. In such situations it is very difficult for anyone, no matter how experienced, no matter how well qualified, to divorce the action from the person, or to see, when they are in the midst of it, that it is a manifestation of an illness rather than voluntary and premeditated action to do harm. In other words it is easy to forget that it is the illness that is causing the action, not the person behind that illness.

Despite the stresses and strains of supporting patients manifesting the extreme ends of BPSD very few – I cannot say none as there are some lamentable exceptions to this rule – CW's I know would advocate the kind of blanket prescription of powerful antipsychotics - as described above - in order to make the job easier – the last thing someone caring for the elderly should want is no interaction between them and the people they are caring for, for this is where the magic of care happens – the beautiful relationships you form, the trust you build, the people you meet – however, equally, in a difficult work environment it can be understandable that care staff either lose focus on the fact that they are caring for seriously mentally ill people and/ or will question why such challenging behavioural symptoms are not being addressed through medication. The question is a difficult one and to show how difficult and nuanced it is, I would offer two examples to demonstrate this.

One resident I supported was what may be described as "a character", they had led an active and intensely interesting life that included service in the army overseas and then a life down the pit. They had been a great dancer and often, when on the wander, hopped out a few steps. In many respects they were a delight to know. It was a different matter though when any personal care was attempted. He might, if you were lucky, allow you to wash and shave him, but any attempt to try to clean his more personal areas (i.e. anything below the waist and above the thighs) was next to impossible, on trying to do this he would get very defensive and then quite violent. As a result, and because of the policy to walk away from any challenging behaviour, he was sometimes left for days, and, on one occasion, weeks, without have *any personal care whatever.* Frequently he walked about in an unkempt state – dirty clothes, food on his face, unshaved, a smell of faeces clinging to him and hair greasy. His family were, unsurprisingly, unhappy with this state of affairs; they were not alone, so was I. Appalled that I was complicit in this care I tried on many occasions to try to get him in some sort of decent shape.

I tried as many techniques to achieve this as I knew, and that had worked in the past on other dementia patients – first was trying to prepare him for a wash by making it a collaborative venture - telling him that he needed a wash because he had not had one for days and, although it would not be immediately, to know that in, say, half an hour, I would return and *together* we would get him washed.

No dice.

Next I attempted to give him little warning - to say that he needed a wash right now and that, again, together we would get him looking and smelling nice.

No joy.

Finally I tried to catch him on waking up – walking into his room with a bright and happy smile and breezy manner and saying it would be great if he could get a wash this morning before he got dressed.

The worst of the bunch – he told me – and I quote – to "F**k off you daft c**t".

One time I came on shift to find him with his pants round his ankles and with a lump of faeces in his bottom. Being fully mobile he was walking around, sitting on other resident's chairs, beds – and worst of all – tables. It doesn't take a healthcare professional to see how dangerous this was in terms of cross infection of any number of potentially serious conditions. Another CA and I made a determined effort to get him clean that involved us in a very challenging situation. We eventually got him somewhere near clean but only after a titanic struggle.

Later I was severely reprimanded for not walking away and leaving him once he became challenging. I admit I may well have been wrong, but at that time and in that situation what was I supposed to do?

This though is not the point, although it is a point I shall return to later; in my time supporting this resident I got inklings of why he was so resistant, particularly to having his trousers removed and any washing of his genital area. On one occasion when reasoning with him as to why it was important for him to get properly washed (another one of my doomed techniques) and finally moving to encourage him to remove his trousers he became very agitated and said –

"I know what you're after. No way, you're not having *that* [his emphasis] with me again. I know how your lot operate. Never again."

On that instance I immediately left the room, I was in no doubt just why he did not want to remove his trousers.

Not being a psychologist I'm in no position to make a judgement but – if I may make so bold for just a CW – I would say his resistance to having his clothes removed was based on a past sexually abusive incident; if I was right this would explain all his behaviours associated with personal care.

I recorded this conversation in documentation and made a point in telling the senior nurse about this the next day. Their response was somewhat discouraging. When I mentioned that the root cause of most of this residents challenging behaviour may be buried in his past and based on sexual abuse they responded

"Yeah probably" and went about their business – totally disregarding anything of what I had said - and this was from a mental health specialised nurse.

No psychologist or psychiatrist was called for, no assessment asked, no attempt was made to inquire gently with his family if they had ever heard something similar. It appeared that just because he was old and slipping gradually away in dementia that they were unconcerned about what may have laid hidden for so long deep in his memory and only now was being uncovered through dementia regression.

It may be asked what this has to do with medication. The answer is - maybe nothing; but if this possible past abuse was now coming back to haunt this resident might the answer be found in a full psychiatric or psychological assessment and maybe some appropriate intervention, psychological or pharmacological. I don't know; what I do know is that issues around personal care deteriorated regarding this resident. Where once they had gone days without a shower or proper wash now it sometimes stretched into weeks; he became even more challenging about even removing any of his clothes or even brushing his hair. In short – and again this is just a personal impression and echoes what I said in the previous chapter – we as CW's were asked to try to care for someone with potentially deep psychological issues that went beyond "merely" dementia and without the resources to do so.

Another example, in fact the example that brought me to look at medication in the first place, was one that I found both the most distressing - in terms of my total impotence as CW and as to the shortcomings of medication for elderly as I perceived them - and also the most frustrating in terms in how clinicians remain remote and even indifferent to situations *as they are developing.*

An elderly lady of 86 was placed in EMI care at a home in which I worked. Although mobile and capable in many personal respects – she could visit the toilet independently with prompts as to where it was located, could feed herself, had good communication skills and capacity in most areas of choice - she was very confused; not only did she not have a sense of where she was, she was also disorientated over her history – for example – and this was a major issue that was identified quickly by night shift staff – she did not recall that her husband was not only sadly dead but that he had died suddenly and tragically, some 20 years ago.

She presented in a constantly fretful state – unsure as her surroundings, seeking reassurance constantly and wandering in a very tremulous and fretful fashion.

In common with many new admissions she constantly wanted to know when she would be "going home" and walked the corridors of the home with her handbag over she shoulder and her coat on. At night it occasionally happened that she would wake up looking for her husband, asking where he was and, most poignantly - and difficult for me when I was faced with the situation – asking if he was dead; even through her confusion she was more than dimly aware that her husband had met a fate unimaginable to those who have not experienced sudden loss. Also at night – logically - in not identifying her situation and surroundings – she wanted to know who CW's were who went to her assistance and why they were "in her home" - this often produced screaming fits and cries of alarm that took a lot of time and care to calm her down from.

It was noticeable that in the weeks following her admission her mental state deteriorated rapidly – she became even more confused, even more fretful, became incontinent of urine, and now woke every night looking for her husband, or soaking wet or suffering from assorted pains that she could not directly locate but which appeared to be giving her some very real discomforting issues.

As CW's we did our best to address her mental state, we explained that she was in a safe place where she could be looked after, that she had nothing to fear and that we would be there for her no matter what time of day – arguably this made things worse as she wondered who these people were who had apparently taken up residence in "her house". Not surprisingly then all this was often to little noticeable avail, she remained constantly anxious, afraid and fearful and now was frequently reduced to screaming fits because she could not comprehend her situation.

Although many elderly have a great deal of difficulty when moving into residential care this was something else on a different scale entirely. All the CW's reported that this lady was stretching our capabilities past breaking point as there was little we could to either clam her down or to ease her distress, it was only then that we learnt that this particular resident had been on anti-depressants, tranquillisers and anti-psychotics *before entering the home* to treat longstanding mental issues, most of them surrounding her inescapable grief at the sudden loss of her husband. On entering the home all but the anti-depressants had been stopped, and even these put on a reduced dose. Is it any wonder then that this lady quickly deteriorated?

When we asked why all this medication had been suddenly stopped after so many years of having taken it we were told that it was the local Primary Care Trust (PCT) policy that all tranquillisers and antipsychotics should be stopped for the elderly once they entered residential care.

The reason?

The over medication of the elderly on tranquillisers and antipsychotics. Like all loopy ideas this one sprang from good intensions that had become perverted in practice. As already noted -

"Reducing the use of antipsychotic drugs for people with dementia is a national priority in England"

In order to meet this target the local PCT mandated that all homes receiving referrals from them should cut out the antipsychotics. The purpose of this target was to prevent a slide back to the bad old days and the usage of the "chemical cosh" to keep residents pliant, to reduce workload on staff and therefore to allow staffing levels to fall.

All very laudable – but it raises two very difficult and obvious questions – the first is if a medication regime has been worked out, and shown to be effective, for someone when they were in their own home *and before the onset of dementia*, then it was presumably with their own informed consent, therefore why stop it immediately *without their consent* when they are moved to a residential facility?

Further, how wise is it to completely stop all antipsychotics just at the point when elderly and vulnerable people are experiencing one of the most traumatic ordeals in their lifetimes – their loss of home, external life and full independence, and then roomed cheek by jowl with other dementia sufferers?

The second question returns to what we have already discussed at length but is worth restating in the light of this particular situation – how are CW's - who know only a little about someone entering a home and therefore are trying to build a relationship "from cold", meant to bridge the psychological gap that medication has left?

CW's are there to try to make dementia sufferers as at ease and comfortable as possible, we are not psychologically trained. In this instance how were we supposed to handle matters - such as this ladies' deep and probably irreversible grief, her disorientation as to surroundings leading to traumatic screams for help and a constant anticipation that at last, on *that* day (which was everyday) she believed she going home and was constantly asking where her taxi was?

How were we to deal with all these complex factors without the aid of the medication that this lady had relied - and for all we know become dependent - on, for years? The answer was we could not and cannot in such instances; instead we were left to try to cobble together ad-hoc answers to this lady's cries for help.

It was obvious though to everyone that this lady was in constant distress and that it was beyond our reaching out to help her in any meaningful way.

Eventually one brave CW ventured to ask if there was to be any assessment of this lady's psychological state and possibly any medication intervention. The answer from nursing staff ranged from – that's none of your business – to – we're looking in to it. With either answer it mattered little as it was not indicative of any purposeful action; the result remained that this lady continued in such a terrible state for months, agonised by internal demons that she was only rid of when her voice too was taken from her by her advancing disease.

Hopefully both these cases will illustrate how medication, while not having the dominant role in care, should be considered to have *some* role in it. It will be argued that both these examples are highly partial, containing as they do deep psychological issues beside dementia, but they are far from outliers. As one of the characteristics of dementia is regression into the past it is almost bound to raise issues long buried in denial or trauma, yet it appears that psychological remedial action is lacking - where it would be available to others - *simply because these sufferers are old.*

Understandably there are real and potent challenges for any psychiatrist looking to fathom what causes lie at the roots of such manifested behaviour, first among them not being able to discuss such mental issues with the sufferer in any meaningful and clinically insightful way; less understandably however is the apparent reluctance of the mental health establishment to address *any of the issues at all.* Equally the withdrawal of all antipsychotics can, in many cases that I have witnessed, have more baleful effects than their judicious use.

As a CW I have no wish to come on shift to find a room full of zombies sleeping away in a drugged up stupor the last years of their lives, however neither do I want to see residents in such deep distress that I can do nothing about it. The Alzheimer's Society notes that the total eradication of antipsychotics is not their aim as -

Antipsychotic drugs ....are....commonly prescribed for behavioural and psychological symptoms in dementia [BPSD]. This is because in some cases they can eliminate or reduce the intensity of psychotic symptoms, such as delusions and hallucinations, and can have a calming and sedative effect............ Drug trials have shown that, for people with Alzheimer's disease, antipsychotic drugs can have a small but significant beneficial effect on aggression and, to a lesser extent, psychosis (delusions and hallucinations).

http://www.alzheimers.org.uk/site/scripts/documents_info. php?documentID=110

it does note, importantly, though that –

"These [beneficial] effects are seen when antipsychotic drugs are taken for a period of 6-12 weeks. The benefits of these drugs for other symptoms and when used for longer than 12 weeks are very limited."

Most CA's would wholehearted agree with this. In severe cases of BPSD what staff want, and sometimes desperately need, is a period of time in which they can deliver care as best they can and eventually form a bond that will replace any drug therapy over time; what CW's cannot do is form such a bond with residents who are either very aggressive or very frightened.

Further, while The Alzheimer's Society discusses the use of antipsychotics in the particular field of dementia care it makes no comment – because this is beyond their immediate purview and interest – about the elderly who already have deep psychological issues. This is an important distinction as made above – some of the elderly are on medication that *pre-existed their dementia* and therefore a full withdrawal simply on age grounds or change of living circumstances seems to be not just wrong-headed but cruel. As is usual in the Care Services a one size fits all policy with good intentions frequently produces perverse results – an aspiration – to reduce injudicious use of antipsychotics - becomes a target and a target becomes an overriding imperative that bucks roughshod over the discretion of doctors and other healthcare professionals. Even in the grey areas we have noted, where dementia and metal imbalance overlap, surely any therapy – pharmaceutical or other – is worth trying, even if it is only to discount its ultimate use.

The obvious becomes apparent here – what is lacking is not really anything to do with medication at all but in the constant monitoring of the elderly in residential care. Too often the elderly considered too at risk – to themselves or others – of being left in their own homes find themselves packed off to residential care and there active and responsive physician care ceases – this is why the elderly have been left on antipsychotics for far to long in the past and why now they are prescribed them hardly at all.

Yet maybe antipsychotics, even in the apparently most urgent of cases – that is in those individuals suffering from the extreme ends of BPSD – could be unnecessary and could be replaced by an eminently less brutal and arguably more effective drug that has the added advantage that hardly anyone would object to its use, and would tackle often unresolved pain issues that are sometimes also attendant with the advancing of dementia and some of the other conditions prevailing in the elderly (see below).

Bearing in mind the consistently reported over – or inappropriate – medication of the elderly with antipsychotics medical researchers were drawn to the subject not of how to treat BPSD but some of the possibly unknown or as yet unexplored reasons for it. As the Nursing Times reports –

"Researchers from Kings College, London, and Norway believed that agitation [and aggression of BPSD] may be linked with pain, and that dementia patients were incapable of expressing it in any other way.

They conducted a study into their theory with 352 participants in nursing homes in Norway suffering from moderate to severe dementia.

Painkillers were issued to half of patients during every meal, while the other half continued receiving their usual treatments."

they found that -

"After eight weeks, agitation symptoms among the group who were given painkillers dropped by 17%. The improvement level was higher than would have been expected from antipsychotic treatment."
http://www.nursingtimes.net/nursing-practice/clinical-zones/older-people/painkillers-could-treat-psychotic-dementia-symptoms/5032692.article

The research and its findings were welcomed by Alzheimer's Research UK who responded by saying –

"This valuable research could boost efforts to drive down prescription of potentially dangerous antipsychotic drugs for people with dementia."

And the government also welcomed it as another string to their bow in the national priority of reducing the number of dementia sufferers on antipsychotics and their replacement with less aggressive drugs by saying –

"This study adds to the evidence in this area."

http://www.webmd.boots.com/alzheimers/news/20110718/painkillers-may-help-dementia-symptoms

This research was released in mid-2011 yet at the time of writing I have yet to see any deployment of paracetamol as a prophylactic medication for those suffering from all the classic cases of BPSD.
Why?
It surely cannot be because of the potential health risks and side effects associated with long term and regular paracetamol use can it? Especially considering that antipsychotics have, according to experts -

"a strong sedative effect [that] can cause dementia symptoms to worsen and even lead to a rise in the risk of stroke and death."

http://www.nursingtimes.net/nursing-practice/clinical-zones/older-people/painkillers-could-treat-psychotic-dementia-symptoms/5032692.article

Yet when I asked a nurse who actually – shock horror – treated my tentative enquiries with interest and spoke to me as a sentient and interested adult - as to why this research had not been converted into practice - this was exactly the answer I was given – that is that the potential for liver damage from regular paracetamol use was the reason why this research sank without a trace.
On face value this appeared ridiculous to me, for it is not only carers who have to cope with the effects of BPSD, it is also the sufferers. Research has shown that dementia sufferers – right through to, and

including, end stage dementia – have moments, sometimes hours or even days, of lucidity where they are acutely aware as anyone else of exactly their position and plight. I have witnessed such cases first hand; in such situations dementia sufferers can be suddenly aware of how inappropriate, abnormal and – lets be brutal here – deranged – their behaviour is or has been. This is heart-breaking to witness – even more so for the fact that such moments of lucidity are frequently accompanied by the most bitter remorse and shame of their prior actions and horror at their present predicament.

One service user I supported was so often so sexually inappropriate that he needed to be monitored as to his whereabouts every minute of every day, he also was prone to outbursts of violence that, although they were rarely harmful to carers, placed himself in a great deal of danger due to his actions (the potential for self-injury from flailing arms or a fall due to lack of balance from kicking legs were acute at such times) and the possible reactions of other service users.

Just as often though he would be suddenly aware of what he had just done and then be suffused with the most profound, effusive and fulsome of apologies; in fact his very apologies themselves became an issue - due to the occupation of time calming him down and telling him that his apology was accepted and that there would be no consequence as to his prior actions and indeed it was forgotten about already – something which sometimes stretched into hours (imagine a whole 2 hours of someone repeatedly approaching you and engaging your full attention with constant acts and words of contrition – while you have a dozen other service users needing attention – very trying).

Although the sexualised behaviour may not have responded to the regular use of paracetamol I certainly believe that his violent outbursts were, as these explosions were preceded or succeeded by his communicating that he was in pain or discomfort in a certain area.

Once more I passed on these pain issues to the nursing staff who, as expanded on below – rarely responded either in a timely manner or at all. I remain in the belief though that if the heightened mood of the violent outbursts could be ameliorated by frequent pre-emptive pain relief then his sexualised behaviour may have been moderated too.

This may seem like poppycock and be pooh-poohed by clinicians but if the heightened mood preceding violent outbursts could be moderated then so too could the heightened mood that often accompanied his sexualised behaviour; at the very least I believe that not to try to test this theory, even just to prove it wrong, did this service users and those at risk around him a chronic disservice.

But all this points to the question of how are carers supposed to manage the extremes of BPSD? If not even a clinically proven low risk medication – one that can be brought over the counter and is used by the general population every day - or a combination of medication and pro-active care, then what?

The answer is simple – leave it to the carers.

Once more we return to a familiar theme, not only does government, the clinical care establishment, and society in general turn its back (for all that people like to think of themselves as enlightened, dementia still is frightening and "unsightly" to the general population and so like to see it shut up in hospitals and care homes, hiding under the pretence that they can be better cared for here, when in fact it is akin to in-prisoning the elderly mentally ill, not to protect them or society but for society to protect itself from something it does not like to face or see) but so too do Care Home owners.

This last group is especially culpable in this instance, for they know that without medication the full weight of dealing with BPSD will fall on their staff yet Care Homes remain understaffed and such staff undertrained *in the light that CW's are being asked to fill the gap left by medication withdrawal.*

If the aim of government and clinicians is to improve the lot of dementia sufferers then they need staff in Care Homes and hospitals in sufficient numbers and with the in-depth training – not just a scanty single "dementia awareness" course once a year - sufficient to meet the needs of often challenging individuals. The fact that this is not the case leaves elderly care sadly lacking in meeting the deep psychological needs that often traumatise the last years of individual's lives.

But antipsychotics are not the only problems that arise from a lack of responsive clinical pharmaceutical treatment of the elderly in residential care. If antipsychotics admittedly create a complex and morally confusing arena, this would appear as rocket science compared to the inability of clinicians to manage the most basic of drug treatments – that being pain relief.

One of the dimmest and least savoury of area in care is the habitual leaving of the elderly in pain and discomfort for want of proper monitoring and dispensing of pain killers. Time and time again I have reported to nurses on duty that Resident X or Resident Y appears to be in some discomfort only to find their response lackadaisical at best or completely uninterested at worst, either taking an eternity to dispense desperately needed pain relief or not doing so at all.

Once more the absent minded doling out of painkillers in the past has led to closer monitoring and stricter guidelines for the dispensing of analgesics today, most especially stronger types such as codeine and morphine. But this does little to explain the tardy or absent response of clinicians to dispensing even simple paracetamol for someone's clearly expressed pain.

There are obviously barriers to the right use of analgesics' not least the barriers in communication of pain suffered between sufferer and carer/nurse/doctor. Frequently dementia sufferers cannot make clear where or how they are in discomfort and that often the only indications that this is so can be either so subtle as to be missed or so violently direct as to be misinterpreted.

Challenging behaviour – as noted above - is one the clear key indicators of an individual suffering pain that every carer in the dementia field is made aware of almost from the get-go. I have witnessed this on many occasions but when I have ventured to suggest to the nurse on hand that the behaviour may be down to pain I have been told to go away and consider my position relative to theirs – that is, sod off and don't presume to tell me my business.

This has happened countless times – in fact so many times that toward the end of my career I no longer shared any thoughts of *why* someone may be acting the way they were. Slapped down on so many occasions I resorted to just plain reporting what I observed without any gloss. If you find this hard to believe I'll go you one better, there have been occasions when I have reported a resident as looking and appearing very unwell and that the nurse may consider calling a doctor, only to find this inquiry refused for no other reason than that I had the temerity to suggest a clinical action.

Before you think this plain paranoia take the following example.

A resident with a known serious blood disorder (myodispylasia) who was normally an early riser and who had a good appetite was one morning both listless and uninterested in getting up, they were pale and withdrawn where normally they were colourful and noisy and, what perhaps was most telling, they could not mobilise having lost all sense of balance. I told the nurse coming on shift that I thought this resident was ill and maybe a doctor should be asked to call. When I came in for my shift the next day I found that this resident had had nothing to eat or drink and had not been out of bed all day. The day after and another shift – again nothing had been done to address the still deteriorating condition of this resident. I again – this time I almost pleaded – approached the nurse on duty to call a GP or out of hours doctor. My next shift – still no doctor.

Finally I gave up – on handover this time I merely said that they were the "same as yesterday"; when I came in that night not only had a GP been but the resident, by now deteriorating rapidly, had been blue-lit to hospital with a seriously low blood count, and - not surprisingly given they had now gone 72 hours with barely any fluids – dehydration. Luckily they survived this encounter. The lesson I finally learnt that if you want a nurse to do something suggest the opposite.

This tale though has an interesting coda –

On another shift a couple of months later I noticed the same resident showing exactly the same signs and symptoms. I reported to the nurse coming on duty that this resident was presenting in exactly the same way as when they were last rushed to hospital. The nurse nodded without reply.

Nothing was done proactively to see if what I had observed had any factual basis, no doctor was called and as far as I know – having gone off shift – no close monitoring was undertaken. Later that day the resident suffered a nosebleed that would not stop, this was followed by an even swifter decline, again the result was that after a couple more hours they were blue-lit to hospital. This time when they were discharged it was to receive only palliative care – they died a week or so later.

As it happened this was a very sick man, however the point still stands that as a care worker nothing you observe seems to carry any weight, what is expected of you is to wash, dress feed, clean, bathe, shower and manage residents as best you can – as for making any observations or – heaven forbid – make any suggestions to clinicians – forget about it.

In this I am not claiming to be a better than average carer, I am only using my own experiences to echo the complains of dozens of carers I worked with who also picked up on signs that residents they cared for were in pain or mental trauma, yet they too who followed these observation up by passing them on to a nurse travelled down the same path of obstructionism and stonewalling I experienced. Carers are not so much ignored as positively acted against.

Yet this is not the worst example of inadequate pain management or monitoring as another trumped even this one.

One Resident I supported was in end of life stage; although not on the dreaded Liverpool Care Pathway (LCP – a decent strategy fallen on hard times because of misuse rather than a fault in the underlying programme) they had been written up for powerful sedatives and painkillers. During handover the nurse going off shift informed the night nurse that there was another dose of such medication ready to be given - as he was due another dose - if the resident showed the least sign of discomfort. When I went up later in the evening to the resident's room to position change - him to avoid bedsores - with his main carer for that night, he literally screamed in pain when we tried to move him in the least. I asked his main carer whether he had been given another shot of pain relief and the answer was no. I actually went to the nurse and told them specifically that this man was in clear pain, and a great deal of it. The nurse sauntered into the room, glanced at the resident, whose face was still screwed up in agony and said –

"I'll give it to him later."

Of course later never came – and I had to help move this resident 3 more times that night, each time with the same result – the hoarse cries shrieking out of him at the merest touch. To say that when I heard this resident had died the next day I thought it – more than any other resident I attended to in their last hours – a merciful release.

Not all inadequate medication for the elderly that I cared for was the fault of clinicians, sometimes the families themselves objected to medication withheld or prescribed *that did not suit them,* regardless of any positive or negative effect it was having on the service user themselves.

One example not only demonstrates this but also illustrates succinctly the nexus that sometimes occurs in pain relief and psychoactive drugs.

One person I helped to care for (I was not his main carer) was 96 and and a World War II veteran; at the stage I came to know him when he was admitted to the home I worked in he had advanced bowl cancer and was, consequently, in poor health. Having had half his bowel removed he was totally incontinent and regularly needed attention and cleaning for bowel incontinence (he had been catheterised for urine retention). Despite his mounting health problems and clear embarrassment as his continence issues he was one of the most stoic, ineffably polite – even when weakened he *always* stood when a women entered the room – and interesting individuals I met in my job. He always showed good humour whatever his situation and was always ready for a chat.

Over time, inevitably, considering his age and health, his condition declined. He began to show signs of constant physical discomfort which was hardly helped by 2 grade 2 bedsores on his lower back (sacrum) caused by some shifts failing to adequately clean him or – at times – to *clean him at all,* and which remained undressed. Shift after shift I worked with this resident they showed all the signs of being in ever increasing discomfort – he could hardly sit or lie and changing positions made him wince; yet, despite having a range of painkillers available for him, never once did I see them dispensed. Equally as disturbing was the fact that this person was in almost perpetual mental anguish – all day - and more especially at night - he would shout for carers' attention –

"Nurse! Nurse!"

In a raw hoarse voice that was visceral in its imploring lamenting tone. Every shift I worked his main carer (he was blessed in having an active and conscientious night carer) would go to his attention promptly but found no object to his beseeching cries – he could not name his legion of encroaching fears nor find respite from them in human company and attendance, neither could he find any object for his agitation, only the fact that he was tormented.

At one point – the mental turmoil being all too evident - he had been prescribed a very low dose of diazepam which, for a few weeks, gave him respite, until it was stopped *at the families request* because he was oversleeping. I saw this man regularly during the period he was taking the tranquillisers and never in all that time was he in a drugged stupor, in fact whilst on the medication he showed signs of being more like his old self. Nevertheless the family disliked that he sometimes slept in until late morning – their complaint being he had always been an early riser - or was drowsy during the times they visited on afternoons, (a 96 year old who slept longer than they did when they were working or when they were physically in better health and who sometimes had a doze in the afternoons – surely not!) so the medication was stopped and within a week he was back to where he had been – constantly agitated and distressed.

Although not a clinician I am in no doubt that this man's last weeks, or months were made unnecessarily painful and uncomfortable by a choice *made for him* by family members *in their own interests, not his.* The clinicians themselves were weak in not being sufficiently forceful in standing up for this residents best interests against those of his family – a dereliction of duty of care that cannot be excused, as they, not the family, were medically trained.

However the indignities were not quite complete - in the last hours of his life – when he was in palliative care - he was put on the LCP and sedated because of his pain and distress, he was dying but he was not dying in pain or torment, however the family requested that he be taken off all sedatives "so they could be with him at the end".

He died wracked with agony and in a place no one – not even the best of carers - could reach.

Quite what the family were thinking I do not know – nor - despite the fact that I believe that their actions constituted a betrayal of trust and a reckless disregard for their kin's best interests – do I wish to cast aspersions or judgements on them - impending grief, separation and loss can do strange things to people – I remember when my own father was dying asking the Macmillan nurse - cryptically - I thought, but painfully transparent in fact – if he could not in some way be "speeded on his way" (I parenthesise not to quote but to illustrate euphemism) only to be told politely that they didn't do that. Looking back the nurse must have thought I was off my rocker, although she probably faced the same questions from most families she interacted with in the final distressing hours and days of life – so I cannot be too hypocritical. However it is also my undoubted belief that this man spent his last weeks in unnecessary mental and physical anguish due to the actions (or inaction) of clinicians and their over-accommodation of a families' misguided wishes.

Not all blame can be said to be borne by in-house nursing staff, sometimes the NHS leaves the private care sector with the proverbial bag of spanners in dealing with an individual's pain relief.

A classic case was where an 86 year old service user and dementia sufferer fell in one of the care homes I worked in, he was taken into hospital where a broken hip was diagnosed. Within a week he was back in the care home, quite why his stay was so short was a mystery but can be explained by the generally hostile mood of the NHS to the elderly mentally ill - they turn them around as fast as they can and wherever possible kick them out in short order.

What was more apposite to us, who had to care for this man, was not so much his seemingly premature discharge as his pain management. In hospital he had been on regular doses of morphine to manage his considerable pain, yet when he was discharged to us he was left with simple paracetamol. To say this was inadequate to manage his pain would be to do violence to language. Over the next 6 weeks or so his situation remained one of constant pain not helped by bedsores on his bottom that meant he had to be regularly turned so inciting even more acute pain.

Because this individual displayed very challenging features to his behaviour it is possible that he had been inappropriately prescribed morphine in hospital so that he would be kept pretty much in a narcotised state to pre-empt his challenging behaviour, but this is, to an extent, beside the point, that being that we were meant to care for an already challenging individual in almost constant unmitigated pain – a recipe that eventuated in several CW's refusing to care for this man because his behaviour became so anti-social and challenging (without dealing in specifics the behaviour was such that we as non-psychiatrically trained CW's – as noted at length in the previous chapter – were completely unequipped to care effectively for him).

However under-medication or the reluctance to use prophylactic medications experimentally proved to work is only one extreme, in one other key area over-medication is almost universal – and that is in the use of laxatives.

In one sense the use of laxatives are emblematic of the problems I have identified above about the inflexibility and unresponsive clinical care of the elderly. Nearly every resident that was admitted into residential care from hospital in my time working there were on quite powerful laxatives. These, in main, were continued, sometimes for the sum total of their stay (which, without being indelicate, mostly meant for the rest of their lives) without any review taken as to whether it was suitable that they should still be on them.

The reasons why most admissions were on laxatives in the first place is that most typically they had ended up in hospital after a fall or other serious injury or a psychotic episode; mostly though it was broken bones or other painful conditions that had precipitated their admission to hospital.

Because they were dealing with painful conditions while in hospital the chances are they were on some opiate type medication for pain management - this can range from codeine through to morphine solution. These have a consequent constipating effect; add to this that the elderly in this situation will have been bed-fast and therefore inactive and you have all the ingredients for chronic constipation. The answer – place them on laxatives, and the more powerful the pain management medication, the more powerful, or frequent, the use of laxatives.

No problem there.

However when they were discharged to a care home it was often the case that the more powerful painkillers were withdrawn (see above for an extreme example) – in some cases this was warranted – if, as should be the case on discharge (but often isn't) they have had the causes for their hospitalisation treated and have received some form of rehabilitation – as such they would no longer need such aggressive pain management and what is more they would be able to mobilise to some degree or another.

Still no problems there.

However, while the more aggressive pain management medication was withdrawn and the individual became more active than during their hospitalisation, the often powerful laxatives were frequently left on their medication regime.

Problem.

I'm sure I do not have to stoop to the graphic in order to anticipate what should be obvious in any readers' mind, yet if explanation is needed then it can be summed up as crap – and a lot of it.

The main problem is not the fact that CW's have to deal with the results of this naked Overmedication - any CW worth their name would not object to doing their duty as regards to seeing to any individual's personal care, and that this sometimes can be a messy business; however to see someone constantly passing loose stools, and then to find that they are on daily high doses of laxatives perplexes and frustrates many CW's; and not just in terms of their job, it frustrates them because they see first-hand how distressing and damaging this can be to the individuals they care for.

Imagine yourself fully continent prior to a period of hospitalisation and now totally unable to predict or anticipate any bowel movement, especially at night while in bed; think how embarrassing and degrading this is to have to either call for help or to make vain attempts to clear up the resulting consequences.

Further, imagine how discomforting it is to constantly have loose and watery stools and the chronic stomach pains that often accompany powerful laxatives. The issue is compounded if the individual is now unable to mobilise; constant loose stools have a seriously deleterious effect on the skin of the area where most immobilised individuals spend the overwhelming majority their time applying pressure from their upper bodyweight to – the bottom; in this condition, regardless - and sometimes because of - how often they are cleaned, the risk of developing bedsores or ulcers increases, and if they do result the chances of infection of them due to the constant bowel movements jacks up even more.

Many times I have seen such a scenario unfolding and yet have been powerless to do anything about it apart from treat the effects. In my experience laxatives are the most difficult medication to have withdrawn because the risks of constipation are considered greater than the results of constant loose stools. I would not argue with this clinical decision, yet I would argue with the lack of oversight following them.

Despite reporting endlessly that such and such a service user was suffering from repeated bowel movements, nothing was ever done. Even in cases where nurses have listened to what CW's are telling them and withhold the laxatives from a drug round, the fact that this decision has been taken, and the reasons for it, are almost never handed over to the incoming shift; so while one dose of laxatives is withheld perhaps two or three other doses are continued with the following shift. Once more we return to a familiar theme – nurses not listening to what CW's are telling them and clinical decisions taken without reference to an individual's situation.

However to totally pass the buck onto nurses on this issue is unfair, CW's too have to shoulder some responsibility.

In most care homes (but not all) residents will have a bowel chart on which the frequency and consistency of any bowel movement should be recorded (the consistency is measured according to what is called the Bristol Stool Scale – see http://en.wikipedia.org/wiki/Bristol_stool_scale).

Most carers though simply do not fill this in because they fail to see the urgent requirement for it. Yet as the above has shown an accurate and constant recording of the type and frequency of movements is a major priority. CW's cannot complain that nurses are unwilling to listen to them if they do not contribute to a evidential record of what they are telling them. Worse I have seen CW's deliberately lie about a service users bowel movements in order to try to stop laxatives that are properly prescribed so they do not have to manage the consequences of bowel movements.

Still, even if CW's do properly record bowel movements, this record often goes unchecked by nursing staff or management – this is the curse of all care paperwork and the so called "Paper Exercise" of it (Paper Exercise is the filling in of useless paperwork that no-one ever has reference to).

All this illustrates and offers, I would hope, a corrective to the common perception of care in care homes that the first recourse is always to medication and its inappropriate or disproportionate use – although it should be added that there are still many, many cases – far too many cases - where this remains an issue.

However, there is one final major, and very disturbing issue with all medications – targeted either at mental or physical health issues - administered in care homes and that is, even when they are properly prescribed, it is frequently the case that those to whom it is prescribed still do not receive them.

I have lost count of the times when, cleaning future, such as seats or even tables in rooms, I have found pills that obviously the service user has not taken – either spitting them out, having them dropped or simply dispensed into medication pots and left either out of reach of the intended recipient or no assistance given to the recipient to take them when it is clearly written in their care plan that they need support in this area. Worse, there are times when the nurse simply cannot be bothered giving the medication themselves so hand pots of pills to CW's. Again countless times I have been handed, sometimes 2 or three medication pots and asked to give them to service users despite me having no formal medication training or being qualified to give them.

Here I should make clear *that a failure to ensure medication is taken by service users is an offence if the nurse then marks down on the Medicine Administration Record (MAR) charts that they have taken them.*

Further, medication practice demands that the person dispensing the pills *has to give those pills themselves* – they cannot delegate this task as they have no way of knowing if the service user in question has taken them. What is common practice is that nurses or senior carers *do not ensure that service users have indeed taken the medication* before they sign the MAR charts. I do not have to explain how serious a breach of practice and duty of care, not to say professional ethics, this is, and how catastrophic the results can be. As shifts change and different nurses or Senior care staff rotate, failure to give or to check that the medication has been taken by one shift can have a culminative effect – if several different nurses are as similarly lax – perhaps thinking that if a service user misses one dose of medication it matters little as they will receive it the "next time" - in dispensing medicines the issue can back up, leaving service users without vital medication for days or sometimes even weeks.

Even more worrying is that nurses or Senior carers sometime do not even look with any degree of warranted attention at the MAR charts – that is they do not properly check that service users are receiving *all the drugs they have been prescribed.* If you think this is hyperbole an example will demonstrate.

A new admission to a home I worked in who had no independent mobility was noticed, after a couple of days, to be very chesty, wheezy and often had clear difficulty in breathing to effect. We, as carers, were told to keep this resident propped up in bed and seated upright at all times because the clinical staff believed this resident had a chest infection and even got a prescription written up by the GP (without the GP visiting and merely going from the patient's notes) for antibiotics.

Incredibly no one noticed that on the MAR chart for this resident was clearly marked that they should have an inhaler. No inhaler was in the medication that accompanied the resident into the home and – this may be hard to believe – *not one of 4 different nurses over the course of 2 weeks* picked up the fact that a vital medication was missing.

It took a new starter nurse actually looking at what was written on the MAR chart to see that this service user should have had an inhaler. She then instructed the carers to search this service user's room thoroughly to see if the missing inhaler was in the service user's possession; eventually it was found in the woman's handbag. Had this new nurse not been alert to the fact of this "missing" medication serious health implications could have followed.

Medication and its uses, as hopefully will have been clearly demonstrated above, is, I believe, a growing problem in residential care for the elderly. The failures outlined above appear to be endemic. Yet all medication is supposed to be audited, not just by nurses, but by managers also, but this auditing would put Arthur Anderson in its pomp to shame, so woefully lacking in rigour is it – failures to properly audit medications included such snafus as out of date medications, the wrong medications for the wrong service users being persistently administered unnoticed, controlled drugs not being destroyed or fully accounted for and, perhaps worst of all, medications for some service users running out without being re-ordered leaving service users without often vital drugs for days.

Yet even the most rigorous auditing will not make up the situations where betwixt the medicine trolley and the service user's lips, there are unpardonable slips. There should be no reason for a CW to be finding tablets down the sides of seats, on tables or indeed finding themselves being asked to administer medicines they are not trained to give or which they have not signed out.

A much more broader issue though, and one that arguably is more significant and weighty for how we view elderly care in general, is just how we choose or would like our elderly to be medicated, and is a conversation between residents, their families and health professionals that has been assiduously avoided for far too long.

It is well overdue being had now.

The problems highlighted, as I have encountered them, all revolve around the appropriate, or otherwise, medication of our elderly for physical and psychological illnesses that present themselves day in day out. I have been at pains to point out that no CW with any sense of care and compassion would wish to see medication used *to make their jobs easier*, what they do want is appropriate medication that will *allow them to do the job effectively*. All the training in psychological intervention in the world will not have one iota of value if it is deployed against a condition against which it is bound to have no effect because some form of drug therapy is totally absent. A psychotic episode, deep-seated and long standing mental issues, or some of the more challenging behaviours associated with dementia all need to have a two track approach to their amelioration; to remove medication from this mix is to leave CW's in a situation akin to – to quote one former politician – sending your batsmen to the crease only for them to find, the moment the first balls are bowled, that their bats have been broken before the game.

This has implications not just for how CW's can manage situations between themselves and service users, it also has even more import in how they handle situations between residents.

One of the CW's lesser known obligations is to see that service users are not just safe from harm from themselves but also *from other residents.*

Violence, predatory sexual behaviour or psychological intimidation – intended or not – of one or more resident on one or more others are all things that CW's become accustomed to seeing; again, if they are thrown back on their own scant resources, such as that of keeping certain residents under observation at all times – it is like applying a sticking plaster on a skin tear – at one time or another a risky – potentially very serious - incident is almost bound to follow if one or two CW's are asked to keep tabs on multiple residents; while medication may not be the only answer here, or even the best one, it cannot be simply excluded by some blanket ban that takes no account of individuals' circumstances.

And this is the crux – medication - as regards the elderly with dementia - is one area where an anachronistic "one size fits all" policy of healthcare is not just tolerated but widely accepted. No other section of the population, including other potentially vulnerable groups such as children and those with severe physical or learning disabilities, is treated this way (although it is accepted that abuses do occur in these matters to such groups but never so reductively and collectively); instead medication regimes are frequently calibrated to meet specific symptoms and ends relevant and agreeable to the individual.

Yet with dementia sufferers it is deemed suitable to make judgements based on generalities rather than the specifics. Worse is the fact that relatives are often allowed to make calls regarding medication about which they are manifestly unqualified for and, I would dare to suggest, in any decent civilised system, should have no right to make, no matter how good their intentions.

What we need to ask is what is best for those suffering from dementia – and I would like to stress that this is not just a conversation to be carried out over, above and excluding the dementia sufferer themselves. Although it is not always possible to include dementia sufferers in the usual way it is possible – yes, and time consuming – to work out a medication regime that suits their lifestyle *as it is now*; not as it was, nor as health professionals or relatives believe it to be, but in best interests of the service user and the mental and physical condition they occupy *at the time decisions are made regarding medication.* Further this should be reviewed regularly to take account of the often changing states service users occupy.

Some suffers will indeed be able to communicate their wishes and preferences in one way or another, but for those who cannot trial and error can and will provide answers – the result should be a relatively calm, happy, fulfilling, active and engaged life free of unwarranted pain, mental distress or harm. I believe this is not a pipe dream but is possible, but first we have to get away from the idea that that some medications are bad *per se*, and none are regarded, by default, as good without regard to an individual's informed choice and best interests.

However this idea of a dialogue will be rendered as useless if residents and their families believe that CW's are not holding up their end of the bargain – that is being able to care for their residents in a safe and responsive way that is alive to the difficulties posed by dementia and its symptoms and have the required expertise and training to be able to create a calm and fulfilling environment.

This brings us to how well trained CW's are at managing large number of dementia sufferers.

CHAPTER XVII

Training Days

You would imagine that CW's in dementia care would have a thorough understanding of the illness that they are dealing with and an insight into how a dementia sufferer may see, hear and appreciate the world, yet in my experiences there was little or none of this that proved to be of practical use in my day to day work. Actual dementia training amounted to a morning session called Dementia Awareness which largely covered the main forms of dementia and their symptoms, then an afternoon discussing and being advised on the management of such symptoms and potential actions and interventions that might be of use in practice. Added to this was a short, general and wide-ranging discussion as to how barriers imposed by dementia symptoms – such as communication and expression of commonly desired ends - could be respectively broken down and met.

And that was that.

Worse, sometimes the dementia awareness course was taken by someone with very little direct experience of working with people with dementia.

Sorry?

Yes, and I'll say that again – and just in case you think you misread then I'll put in bold and underline it, what the hell I'll italicise it too - *sometimes the dementia awareness course was taken by someone with very little direct experience of working with people with dementia.*

However unbelievable this is, it was true that on several occasions the person taking the course was a professional speaker *with no experience in care at all never mind dementia care.*

The value of these "Dementia Awareness" session - hobbled from the outset by the simple lack of experience of the course leader – was further degraded to the point of paralysing narcotisation by the fact of the course leader simply read word for word from a hand-out they had distributed at the start of the session and this rote speaking was augmented with PowerPoint slides that *repeated again* what was written down and what was being parroted.

It was not that I was uninterested in learning more about dementia, it's causes, treatments and how those beset by the condition could be best cared for, nor, more to the point, did I approach these sessions in a mindset of having nothing to learn from them – in fact I was desperate to know more. The problem was that the information was disseminated in such a deadening fashion that any potentially useful cues that could be applied to my everyday work was lost in the mind numbing delivery of it.

This drove me instead to learn more on my own account; not through any inherent virtue as an individual carer but simply because I was desperate for a mental paradigm that would enable me to contextualise how I could best meet the needs of often challenging individuals or break down barriers to effective communication between us. The fruits of this will be covered more below.

What I wanted from these sessions in short then was a basis, a background and a knowledge of the types of dementia, their symptoms and possible causes and how they manifested themselves in the actions of sufferers; I wanted a theory on which to ground my practice.

Such courses supplied none of this though, they were simply box ticking exercises by which the home and the managing company could show the CQC that they were providing the "relevant training."

Aside from my own feelings I wondered what someone newly started in care – of which there were usually three or four on any one occasion - with little or no practical experience were supposed to take away from such a sessions, what, not only did they *not* learn, but what kind of example did it set them of the *usefulness of training in general*. At best they would surely think it valueless, at worst they would become inured to learning more than they currently had picked up on the job.

Luckily – as a counterweight to these enthusiasm crippling experiences - on a couple of occasions the Dementia Awareness course was taken by enthusiastic and vastly experienced trainers who either were totally immersed in the vocation of care or, even more importantly, *still worked in dementia care.* On these occasions the discussions were lively and enlightening and yielded some valuable insights into methods and approaches that might be more effective in the care we delivered.

This highlights two central issues with training in the care sector – firstly its quality is hugely variable and can range from the useless to the stimulating, from the stultifying to the vivifying (we will return to the problems this causes shortly); and secondly - and arguably more importantly - the very key area of understanding that CW's need in their job – an understanding of dementia – is covered so sketchy and in so little depth.

The important point here is that while Dementia awareness only covered barely a day, most of the other training concentrated on the *management* of Dementia sufferers; moving and assisting (lamentably termed moving and *handling* in most work places - as if we were dealing with lumps of meat) infection control, health and safety, food hygiene, and other similar topics. Although each of these are important - and I have no wish to downplay their value - it displays an ethos by which each resident is presented as a set of "problems" that needed to be overcome – this over concentration on seeing each individual as a bundle of tasks to be mastered devalues their humanity and creates a mind-set that a day consists of a number of "operations" that have to be performed to produce a desired end.

Where is the individuality in this? Where the dignity? Where the respecting of a person?

It may seem as if I am being overly analytic in this criticism, looking for defects where there are none, or, if any, only minor ones; yet I believe not. If CW's are trained in proportion to what they are supposed to *know* in carrying out their jobs then this upside down ratio is bound to create and similarly upside down view of care as it should be carried out at the point of contact with residents – management first, understanding later.

It may seem a small point, but it's the small points that count.

Returning to the variable quality of the training itself – in whatever area it covered – it singly failed to tackle two major issues that remain a problem in care - firstly is that most carers will pick up how to care from the people they work with rather than from trainers who educate in best practice; and secondly, for those with long careers in care already massed behind them, it fails to properly challenge entrenched attitudes that often go unchecked.

The first point – that most new carers learn from other more experienced carers - comes back to the paucity in inductions into care work, an issue that has been covered at length previously above; proper inductions would see carers educated primarily in the approved methods of working - based on best practice - rather than those prevailing among the workforce - which are largely based on expediency; but training should also offer a corrective to individuals' who have some, or even lengthy, careers in care who may be continuing in outdated, now unapproved practices, or even just plain wrong or even abusive ones.

In the main most training fails on any measure of addressing either of these two issues as most of it ranges from the useless to the esoteric as applied to the actual delivery of care, with perhaps a happy medium only skated over or failing to be reinforced. Because of this an attitude thus prevails that what you hear and maybe (just maybe) learn in training is fine in theory but unworkable in practice.

I myself can claim to be guilty of such statements and sentiments, and although not excusing them I believe that they are the product of excessive work pressures brought on by perilously low staffing – something particularly prevalent in the private sector where I predominately worked in my time in care. To paraphrase the Roman poet Ovid, often I saw and approved better ways but followed worse. Best practice often in care becomes just that – the best way to carry out a task, but in care, with rapidly changing priorities and the constant possibilities of things going awry it is not always, or even mostly, possible to follow recognised best practice.

This is not the same as being insensible or dismissive of best practice, but a recognition that while it may be an aim it is not always or, again, even mostly, possible in the prevailing work conditions.

Yet work pressures and rapidly evolving situations that demand immediate rather than best care responses cannot totally exculpate myself and a vast majority of carers I worked with. Too often I found myself working in a way that would best convenience me and in ways that, while not being detrimental to service users, did not always meet a level of care that I should have delivered. Of these intentional lapses are practices such as using non-approved moving and assisting techniques with respect to service users, not always following usual infection control measures and not always working in pairs with a service user who requires the attention of two carers. Nearly all of these were based on expediency and still this is no excuse. Training sometimes did make me look at what I was doing and so adjust how I worked but this did not happen as often as it should, which brings us to the second issue.

Although I may have departed from the approved way of working and although I was often largely guilty of carrying bad experiences of training into the next session of it I would like to think that throughout my time in care I always remained teachable and open to new and better ways of working. I believe this is not an idle or self-serving claim, as often I went into training with low expectations or a negative attitude only to be surprised and interested in it despite myself. I also always liked working with enthusiastic new carers because they often shamed me in how I worked on a practical or empathetic level and made me re-evaluate how I did my job, but added to this they frequently came up with much better ways of working that both made the life of service users and us as carers, better, happier and easier.

This was rarely the case with those who had worked in care for a long time. One of the most irritating manifestations of those who I would call "old soaks" – those who boasted they had worked for 10, 20, or even 30 years in care - was an arrogant attitude that they could be taught nothing they didn't already know. This arrogance often extended beyond colleagues to that of the trainers themselves. One "old soak" I knew, at the start of one training session and in the hearing of the trainer, leaned back in his chair oozing whatever is the opposite of charisma and declared loudly that they had attended such and such a course so often that –

"I could give the training session"

Even at the time, when I had spent only just over a year in care, I thought this attitude not just arrogant but also thoroughly rude to the trainer present, I also wondered how this carer could be any good if he thought that there was nothing he could be taught (I may have not been impressed with a trainer but only after experience of them – not prior to the training starting).

In my experience anyone – and there were many such – who had this attitude was poor carer; there was no barrier to good care, I found, as high and un-traversable as the wall that declared there was nothing new to learn.

Care, as I saw it, and still see it now, is a constantly evolving workplace; what may have been approved yesterday was today considered not desirable, what may have worked yesterday may not work today, what you knew yesterday you may have to learn all over again today. Without such evolution we would be back to the situation described in the opening of the preceding chapter where the elderly are medicated into submission and kept in a state of docility that robs life of all meaning.

Training should tackle such attitudes of long time carers, the fact that it does not is to some extent not the fault of the course or the trainer, as against such an attitude of implacable professional arrogance there is little that even the best of educators can do. However it is also true that not enough is ever done to challenge such entrenched attitudes, not enough is done to root out such misplaced confidence in experience and ability, not enough is done to show that there is as much to learn at Year 20 as there was at Day One in care. Often a trainer, usually one who has limited experience of care, is intimidated by the windy statements of – "I've worked X –Tens of years in care" and ends up pandering to, rather than challenging, such bloviating clowns.

However even in best of courses taken by the best of trainers in dementia awareness there was never any attempt to go further into just how dementia affects the brain and thus why sufferers act as they do, no attempt to place us as carers in the position of service users through role playing or even discussion of the finer points of how we carers do what we are supposed to do – that is care and its responsive delivery - and very little attempt to challenge perceptions or practices in any real or substantive way.

These missed aims sound very airy and demand some expansion because they go deeper than at first is apparent - into just how care provision for the elderly is delivered and how we as a society treat our elderly.

The first point – the lack of an introduction into how exactly some dementias affect the brain and how this effectively robs care of important context in its provision at the outset we will be come to shortly, so let me turn to the second issue – the role reversal of carers and service users.

Care, despite often the best of intentions, can become a very adversarial environment; simply put, it can boil down to getting service users to do what you want them to do. This is a road bound for hell and opens the way to all sorts of bad practices and often is the lead in to abuse.

The very basis of care is empathy and to empathise truly is to put yourself in someone else's shoes. Yet it is amazing how often, even in daily life, even when dealing with people we know well and, most startlingly, even with people we love dearly, that we singularly fail to see a situation or life itself from someone else's perspective. How often do we arrive at confrontation simply because we cannot see the other person's point of view? Dismissive treatment, careless words, a simple gesture or tone of voice can set get our hackles up, yet we do the same to others every day. As a result we see actions made toward us as consisting in maliciousness and unfounded hostility without seeing them as possibly a reaction against something we have done - we see the consequences but not the precipitory actions underlying them, nor do we perceive that we may have been the cause of them; then in hurt and in a lack of reflection we in turn react upon the reaction – this is the stuff of irrevocable relationship breakdowns.

Equally the opposite may be true - we frequently fail to "read" people and their actions without ultimate recourse to ourselves. Someone may be short with us, they may make a careless comment, they may be defensive in body language (James Borg states that human communication consists of 93 percent body language, while only 7% of communication consists of words themselves, although this high percentage is disputed by Albert Mehrabian who states "that the verbal component of a face-to-face conversation is less than 35% and that over 65% of communication is done non-verbally" http://en.wikipedia.org/wiki/Body language) and we see these as directed in their totality toward us, instead of us considering possible reasons for these actions external to us – such as the person in question may just be having a bad day, they may just have had an argument with someone else without our knowledge, they may be stressed or under pressure – we instead internalise it an believe we are the focus of it. Such egoism verging on solipsism is stunning to us when isolated like this; yet we formulate these reactions every day and in almost every relationship we have.

In brief the complexity of human interaction is endless, convoluted and takes wonders of cognition to navigate it, yet in the attempt we often get things wrong.

If, then, we so often misread (or simply fail to read) mistake and misinterpret others' actions and reactions - who are familiar to us and in their "right mind" what chance do we have for those who are not in possession of all their faculties? In working with service users suffering from dementia it is almost impossible to read signals accurately, in fact it is almost impossible to decipher the differences between what is a reaction toward us as CW's and the way we work – something we have control over – and what is a reaction to the living circumstance that those in residential care find themselves in or what is behaviour subject to elements of their disease - both of which are beyond our influence.

Most training emphasises the importance of the above element that CW's have control over – that is our personal presentation toward service users of verbal communication, body language and non-verbal cues and communication signals and how this presentation should guide the way we deliver care; but much of this is of the sort that it wholly didactic – that is – "you must do this, this and this because it is the right thing to do" it never asks the question of CW's why; why should we act this way? Simply because we are told so? Because not to do so would result in censure? Because it's part of the job tasks?

This isn't learning, this is the application of mindless authoritarian discipline without a justificational foundation and ultimately leads to CW's avoiding these imperatives when and if they can because they have no practical basis. The fact of how - by checking how we as CW's work, act and present ourselves – will contribute to a more harmonious work atmosphere, will cut down on challenging behaviour, and will make the service users' lives better and our care more effective is almost never covered.

Moreover, it teaches absolutely nothing about empathy skills because it never places us in the position of service users. This is to put not just a cart before a horse but to crash a veritable fleet of pantechnicons right on the nose of a knackered donkey – instead empathy should be taught first and then the other elements of how CW's present themselves to service users and then go on deliver care should fall into place through this insight, otherwise – as is the case now – CW's will attempt communicate with, and care for, service users in the accepted best practice fashion *not because it is simply right and good in itself but because to do otherwise would lead them into trouble with supervisors and management.*

An obvious issue arises here that has been a constant feature of Serious Case Reviews and inquiries into nearly all the worst cases of neglect and abuse in care in recent years, and this is - can empathy really be taught? Is it not something that should be innate in anyone in the caring profession? Shouldn't it be a basic requirement of care staff?

As to the second two questions, and as the preceding discussion about recruitment and screening of CA's should show, this is obviously not the case; so it becomes even more imperative that at least there is some attempt to challenge the first question and to at least try to teach empathy even if it is practically virtually impossible – better that the almost impossible be tried than for it to be left totally absent. Personally I believe empathy can be taught - first by placing yourself in the position of the service user, and, secondly - by employing models of how we see dementia pathologically that will help us to better understand what service users are going through.

The model of how dementia effects sufferers we will come to shortly, so first we turn to how we can put ourselves in the position of service users. So let us start with imagining the following scenario –

Put yourself in this position – imagine yourself out on a night in a not quite blind drunk state, you barely know where you are but have an idea that you are pretty far from home and in a strange pub where you recognise no one. You vaguely remember leaving the house but have lost large parts of the night, in fact you're not sure if this is the same night you remember leaving the house or if it is another night and you have confused the two, you have a recollection of acting absurdly, even badly or violently but simply have no concept of just what you did, or why indeed you did it, other than you are becoming incapable because your mind simply won't work in the usual way......

Now you're on the streets and all is deserted, you wander aimlessly feeling the cold now and all you want to do is get home, then you finally trip up fall over and its lights out.......

You come round but are still pretty far gone, you are somewhere warm but you know it is not home. People are talking to you, asking questions but you catch only a little of what they are saying and reply to what you *think they are saying*, you repeat that all you want to do is get home, but for some reason you are told that this is not possible. You feel bereft, lost, it is like some horrific movie where you are denied the very thing you want – comfort and security, familiarity and safety.

Now you're now left alone but cannot for some reason get up, when you try someone comes and admonishes you forcefully, the sides of your bed are raised so now you cannot get up. You can't help it but you wet and soil yourself. Two people come in and manhandle you about the bed. Your backside is wiped with a rough towel, you're put in unfamiliar clothes - a smock or something - you have something pushed between your legs that feels like a nappy and some tight fitting underwear is rammed on. You feel sick and thirsty and are given some water. Then you are left alone again. The lights are on and although you would like to sleep you cannot. You shout. You shout again. Nothing. So then you shout some more. Someone else comes, you're given a pill. You feel alone.......

Now you are in an ambulance on a stretcher, the doors are open and all you are wearing is that smock and that nappy - which is now soaked and cold; its daylight, you have a hangover from hell and your brain can make little sense of what is going on. You lye for what seems ages with the wind whipping in the back of the ambulance. Then after seemingly hours the back doors are slammed shut on you and you start to move.

Suddenly you are heaved out the ambulance, through sudden cold open air and then into a warm environment. It smells of urine and disinfectant. Images fly by as you are taken down corridors and up in a lift. You swing through a door that locks behind you and are deposited in a place filled with lots of shouting and hostile people, you are wheeled into a room and hoisted out the trolley and into a bed, the room around you looks nice – but it isn't home. You feel drunk again and zone out.

Now you are being shaken and someone is speaking very loudly to you, all you want to do is go back to sleep but they are persistent. You have the cold nappy pad taken off and another dry one put in its place. A cup is jammed into your mouth and you sip some lukewarm tea. You fall asleep again.

Now you are awake, people come into your room, you ask where you are and are told this is your home now, this is where you live. Only you know this isn't home, isn't even close. Where is your house? Where are your loved ones? Where is your music? Where is your television? And where the hell is your phone, i-pad or X-Box?

How do you feel?

How do you want to feel?

Who or how can you get to feel like that?

And most importantly – how would you like to treated?

This chain of events is typically what most elderly go through prior to entering a care home. In such a situation would you like to feel that the people around you try to understand your predicament? Would you like them to try to understand how you feel? Would you like to be treated with sympathy and respect? Would you like your needs to at least appear to be met out of concern for you, not because it is another task that someone will try to fit in between other more "urgent" tasks?

This is the major element played by empathy and its very importance - CW's need to - are duty bound to - understand the people they are supporting and to try to make lives that have been existentially turned upside down begin to make sense again.

To me empathy is like a muscle, it needs exercise to get stronger, and the stronger it gets the more up to the strain that you put it under the better it will be able to take it. The problem is that it is far easier not to exercise it, to let it waste and to use other more brute "muscles" instead.

To show how this can be the case, let me put things from the other side; the CW's side.

You arrive at work to find that 2 CW's have phoned in sick so you half the staff on your unit than you should have, you are then given a short briefing about what has happened over the last 12 hours, most of which is either filled with useless detail, no detail or details that apply not to you as a CW but to nurses – i.e. clinical issues.

You arrive on the floor at 8-00am, you have 15, 20 or even 30 residents to get up. Some service users are wandering about semi-clothed, some of these wanderers are walking in and out of other service users bedrooms, some of them are soiled. Other service users are lying in wet beds, some are challenging - they kick and fight to resist either getting dressed or changed out of a wet bed or clothes – others are very confused and so it makes dressing them very time consuming. Some service users voice the desire not to get up at all but as they have lost weight over the last week you need to get them up so they can have a good breakfast. The breakfast trolley arrives, the chef is anxious to get the food out but you still have another 5 or six residents to get ready, you ask the chef to wait but they have to go to several other units after you so is impatient to get the food dished up, so one of you has to try to get everyone in the dining room and to make sure the right service users get the right breakfast. The other one has to struggle on, getting the remaining residents ready. Everyone is served their breakfast, apart from the ones still missing or those 7 or 8 service users who need support in feeding themselves. The CW in the dining room makes drinks for those present and tries to start feeding one of those who need support.

The manager appears on the floor and starts ranting and raving - why are not all the residents are in the dining room? Why have they not got bibs on? Why is the food is going cold for those still not dressed and why is this or that resident not getting support with their meal? After you have been thoroughly reamed out they turn on their heel and are gone.

One of you, with the now dressed last residents, arrives in the dining room and start feeding those who need support. One of the late arrivals complains that their breakfast is cold so one of you has to go back down to the kitchen to microwave their food. Suddenly a new resident who has arrived from hospital the day before and who you have never supported before heavily soils themselves, the one CW present cannot leave the dining room unattended because of all the potential risks there are there. The other CW arrives with the warmed over food, just as another resident complains their food is cold. The soiled resident now has their hands in their soiled underclothes in a misguided attempt to clean themselves up.......

What do you do? Understand that this resident has, over the last 72 hours, undergone one of the most traumatic experiences of their lives? Tell the resident not to worry? Take them gently, and in their own time, to their room offering words of comfort and kindness, change them with respect to their dignity and individuality, obey every single rule about infection control? Or do you hustle them back to their room strip them while barely conversing with them because you have no time for pleasantries, get them cleaned up as fast as you can and take a flier that you won't be caught as to shortcutting some infection control measures?

That is reality, brutal as it is. The major challenge for all CW's is to maintain a degree of empathy and understanding no matter what the situation is. What is often not understood is the amount of thinking this entails – if empathy is a muscle, thinking is the exercise for it. As in the above example will you as a CW take just a moment to pause, to take yourself out of a task orientated role and *think* how that soiled resident is feeling, *think* how best to treat them, *think* about not just what you have to do but how you are to do it?

Or do you just look at it as another job to be done, and a damn inconvenient one at this very moment, and try to get the situation sorted as quickly as possible.

It is obvious which course is the right one but do you actually take the time to *think* and exercise your empathy?

Care, good care, for all that it is classed as a low skilled job is in fact all about thinking.

Italics, bold and underlining time again – ***<u>Care is all about thinking, and thinking constantly.</u>***

Before doing anything every CW has to think how this resident may be feeling, what are they themselves thinking, what kind of reaction to them is best to put them at ease and comfort. What's more, if that were you, if you had gone through the above chain of events that landed you in residential care, how would like to be treated?

This is a gargantuan task. More often than not at the end of shift I am more mentally tired than physically so, this is because of the time taken always to be alert to trying to understand the people I am supporting – like any exercise, working with empathy takes huge amounts of energy.

Have I lived up to these high aspirations at all times? Always? Sometimes? Frequently not?

In fact in the main I have not.

I can recall any amount of situations when I have failed on almost every count of how I should have reacted with empathy, how I should have *thought* about what I was doing. There were too many times to count when I cleaned up a resident *unthinkingly* as if I were cleaning a blocked drain – a workmanlike approach to an unpleasant task that I sought to complete in the shortest amount of time possible. There were also often times when I and another CW *unthinkingly* carried on a conversation between ourselves while almost completely ignoring the resident that we were supporting with personal care. This is inexcusable, but ask any CW if they have dome similar and if they are being even halfway honest they will reply in the affirmative.

Keeping service users feelings and considerations front and centre simply by *thinking how they may be feeling* is vital, not just because this is the caring thing to do, not just because it makes for a more harmonious relationship, *but because it exercises your empathy.*

The danger, which I alluded to above, and which was present in the poor care mentioned in the immediate above, is to fail, *or to choose not to* exercise empathy and to instead rely on brute task orientation rather a person centred approach. This is all too easy as it feels so much simpler. Therefore just to place yourself in the position of a service user is not enough on its own, some idea of what a service user may be going through as they experience their disease is also important.

We as a generality of people tend to think of dementia as a degenerative disease and see its workings as an attack on the brain that drills holes in it until it resembles Swiss cheese. This in main is wrong, although typical of some organic brain disorders it does not fully explain different types of dementia. For example it took me reading an article in The Economist magazine to find out that most dementias affect the brain not by shutting it down but by making it uncontrollably active – that the brain has no way of regulating its own wildly firing neurons.
http://www.economist.com/news/science-and-technology/21579792-search-treatment-dementia-continues-beta-testing

For me this opened many more doors to understanding dementia that the "Swiss Cheese" model. No longer did I think of someone as *loosing* something but instead I saw them as prey to a mental restlessness similar to a heightened state of elevated experience that leaves them *constantly mentally active but with no focus or direction.*

The analogy I always liked to mentally picture - to put myself in a dementia sufferers shoes *as they experience their dementia symptoms* - was of a situation close to having several urgent tasks to be carried out in desperately short time period – if you place yourself in this position you will surely identify with what I found – that under such conditions your thinking is not clear, you try to do everything at once and so complicate simple tasks, making the whole process more difficult and frustrating, and, what's more - you deeply resent and easily become angry at anyone distracting you or preventing you from carrying out these vital tasks quickly.

Create this mental picture and it is easy to see how someone with dementia can quickly become challenging.

Imagine, if you will, yourself rushing to get several tasks done quickly all the while someone keeps intervening and telling you not to do this, or that that it is not important; throw in the fact that it is not clear at all why the interrupter is preventing you from getting on with these tasks or what alternative solution they may have to your dilemma and you have an explosive mixture. In such a situation I would imagine most people would be tempted to smack the interfering party in the mouth.

Pile on top of this difficulty the fact that the messages that your brain is wildly firing are going to many of the wrong places, like so many signs in a once familiar town that point you in directions that you instinctively feel are leading you further away from your real objective, so much so you have no choice but to act against them – even though in actual fact the signs are right and your instincts wrong.

Again an analogy I like to use is when you may have been diverted off a motorway, say, and seemingly led further out into the wilderness, further from familiarity and even further from getting back to where you want to head. How many of us in that situation have decided to ignore – against all reason – the diversion signs and strike out on our own only to find ourselves seemingly inexplicably even more deeply lost and further from the path to home.

This realisation, this insight, was more use to me than all the Dementia Awareness courses put together, it made me understand why some service users seemed obsessed with getting something done that appeared trivial or impossible – such as making their bed or trying to get dressed in the middle of the night or demanding that you show them the way out because they have to get the shopping in.

Only once I understood that dementia sufferers were not only being driven on by implacable mental imperatives, but that such competing imperatives were making their thinking muddled as they tried to balance them according to an imagined pressing timescale; only when I fully understood that for dementia sufferers what they were pursuing *was entirely rational to them, and* that it was unthinkable that anyone should want to stand in their way; only when I took all of these challenges in, then and only then, could I even begin to truly empathise with what was happening to them.

This better understanding reinforced what I had been told in training courses but had never been told the reasons why.

For example we were often lectured to that we needed to talk slowly and in simple language to service users and to always explain why we were doing what we intended to do, all this without an explanation to us why; why should we do this - other than "so they – the service user – can understand your actions" as if they were a child or a simpleton. Only by bringing my mental picture to this blunt instruction could I eventually understand just why it was important to speak slowly, calmly and simply.

Once again place yourself in the scenario above – the multiple tasks that need doing immediately – then imagine someone intervening – imagine them telling you in a calm and clear way that in fact there is nothing that needs doing urgently, that you, within reason, can do as you wish but that there may be a better time or place to do it. Imagine yourself slowly being put at ease about these competing tasks by an understanding and caring interlocutor between you and your impulses, imagine the pressure suddenly being taken off you, imagine slowly understanding that most of these tasks are either not important or are no longer relevant. How would you respond to this - With relief? With gratitude? With a basis of trust? I would expect all three; all of us at one time or another are happy to have pressure released by someone with perspective, and, although we may not believe them at first, a simple faith will maybe produce a two-way understanding and form a platform from which dialogue is possible.

Yet why wasn't this insight more widely shared even if it maybe scientifically flawed in its actual detail?

I could not understand this. Dementia awareness stressed the central and critical issue of relating to service users on a human level, looking at their backgrounds and histories for clues as to why they may be behaving in an antisocial manner, but failed to impress the added aspect of mental illness on top and overlaying this history. For example a night wanderer who refuses to go to bed or sleep and appears restless, discontent and disturbing to other service users trying to sleep may have worked many years on a night shift so believe that they must be up and about throughout the night as "they are going to work" and cannot understand why so many others are, or wish to be, asleep (this scenario actually happened to me in a service user I supported). Yet a crude history does nothing to help any CW *understand the totality* of the behaviour manifested as not just the crushing effect of a displaced past but as also as applying to the *totality of the mental illness that complicates matters* - by mentally ill here I am not just referring to the loss of awareness that they are now retired and have long since given up work, I am referring to the more complex scenarios that drive behaviours.

An extended look at this "night wanderer scenario" may pay dividends and explain what I mean.

While understanding someone's history is vital, it is useless in actually understanding certain manifestations of behaviour. To use the example above of the night wanderer – if we were dealing with just a misplaced past it would be possible to reason with the service user that while they may feel they must be awake they must also have consideration for others who wish and are able to sleep.

In dementia sufferers this appreciation is not always possible. This can be terribly frustrating for a CW, a service user may be able to understand and grasp most concepts but they may be totally "unreasonable" in this situation; faced with this carers can quickly become irritated or even angry. However if we take the above mental example of the wildly firing brain making demands on consciousness and often leading them in "wrong" directions, their behaviour becomes less mysterious, less frustrating, and a carer therefore can be much more responsive. For example again - the night wanderer may believe he is on his way to work *and* be looking for his mates to go for a drink – two seemingly competing and mutually exclusive motivations that may be present in the service users' consciousness *at the same time.* In this light they are not disturbing people but calling on them to go for a drink.

I have used this extended example because it is only through this process that I am able to demonstrate just how complex apparently simple "challenging" behaviour is – it is not based on one simple cause – there may be multiple causes and they may all exist at the same time making them impervious to reason or explanation. Every CW should know this I believe, it will not deter challenging behaviour, but it will allow its better understanding.

This is nub of many of the Dementia Awareness courses I attended – they mainly dealt with *managing* service users with dementia – this left a huge gap – how can you manage something you do not properly understand? Without being too Freudian let me turn to a simple joke.

There are a series of generic jokes that begin with -

"Why did the chicken cross the road?"

With a suitable punning reply. The joke loses its character if it began –

"The chicken crossed the road."

The point here is unless we ask *why* we are only left with simple facts or behaviours.

Presumptuous as this may be, I believe that all Dementia Awareness courses should include some explanation of what is happening inside the sufferer's brain and thought processes. No recondite elaboration is demanded and not even a model that is entirely accurate (my mental picture constructed above is wildly inaccurate and/or simplified and will no doubt produce howls from any scientist working even on the fringes of dementia research); after all many of us who have studied basic physics or chemistry will be familiar with the "planetary" model of the atom, of a nucleus surrounded by whipping satellites of electrons in stable and calculable "orbits"; only later are were we informed how flawed – indeed how wretchedly factually and scientifically inadequate - this model is; no matter, it serves a purpose, it gives a basic understanding of why matter is as it is.

So too in my above mental picture of what is happening in a dementia sufferers' brain – the fact of whether it makes the grade of scientific exactitude is largely beside the point – if it could be used – as I have found it beneficial – to give someone a an explanation – no matter how flawed – of the symptoms of the disease they are trying to ameliorate then likewise it serves a purpose. In such an instance the finer points of scientific inaccuracy can be said to be largely minor or even redundant as, after all, we are not in the business of training nascent neuroscientists or neurobiologists, we are in the business of understanding and caring, thus anything that contributes to better caring is surely to be welcomed.

And yet, for all the recent focus on care for elderly and those suffering from dementia, for all the reviews into the roles, functions and training of CW's, and for all the fact that a rapidly growing elderly population make these issues more important than ever - care *is still*, at present - and for the foreseeable future - reduced back to the blind management of behaviours of anonymised and depersonalised "problems".

Yet perhaps this is not so perplexing if we look at the structure of care.

Many private care homes are actually not too much interested in having carers who have some understanding of dementia, it is all about management free from understanding, this is because private care stands on two pillars for its continuing endurance – first is, as we have already had cause to mention, the infinite replaceability of carers; and second, the primacy of throughput of paying customers in their homes.

Care homes want trained staff because of grants from government that qualified staff bring, that is why they are so eager to push their employees through such qualifications as NVQ's (National Vocational Qualifications) however they do not seek *informed* carers as these are the most likely to ask awkward questions about the structure of care in the care home sector – questions like why are services so short staffed, rushed, pressed and therefore can offer only scanty care – like why is it that so few service users are occupied through the day with meaningful activities – like why there is rationing in food and continence aids that bear no relation to an individual's needs. In fact they may ask questions that lead to the second point made above – most care homes are effectively battery farming the elderly in which service users are useful so long as the cost of maintaining them is less than the value of the golden eggs they produce in terms of fees.

You may think this an exaggeration but I would ask you to look at the simple facts and, like the chicken crossing the road, ask why. For profit care homes are there to turn a profit as I have already mentioned, companies will look for every possible way to make more money from the service users as possible – this is called a fiduciary duty – the duty to enhance profits by all legal means.

This role will be looked at in the next chapter.

CHAPTER XVIII

Failed State

Now, however, with several decades of experience behind us, the market does not seem all powerful; the private sector alone is not always the answer – *sometimes the state can actually be more effective than business* [My italics] and is a precondition to effective market activity in today's complex world.

Ashraf Ghani and Clare Lockhart - Fixing Failed States.

In their book into why nation states fail Ashraf Ghani and Clare Lockhart identify, as can be imagined, any number of reasons, and it may be wondered just what, say Afghanistan - where Mr Ghani was born and is now President - has in common with private for-profit care homes (to anyone who had worked in one they may well say quiet a lot, with only a hint of irony) however Mr Ghani and Ms Lockhart make some illuminating points which if we tweak the context only little can be seen through the eyes of the what is grotesquely called the "care industry" or "business".

Take for example when they speak of the situation of civil servants who toil away to the best of their ability in dysfunctional countries –

"But the worst problem is that native civil servants are so continuously berated by…. "experts" as incompetent that [they] come to believe this censure……they find themselves maligned by the very [people] they hope to serve as a result of the government's failure to provide even the most elementary means [to allow them to do their jobs effectively]."

In the above quote replace "native civil servants" with Care Workers, "government" with private care management then you have an almost perfect description of care in private sector: CW's are so constantly told that they are not up to the job of caring by outside "experts" that they come to believe that fact, something that is reinforced by the situation whereby they become the lightning rod for all relatives' complaints about care delivery, whether or not Care Workers are at fault; whether or not they can rectify or address those complaints directly.

CW's struggle on against such opprobrium and also against the impossible demands of an often remote management - who make "policy decisions" far removed from the reality of the care environment – and where they are abandoned to carry them out with neither the resources, training or tools to do so, be that in skills or sufficient staff; until, that is - as Mr Ghani and Ms Lockhart note - discouragement and demotivation take hold. The result of this process - as noted later on in their book - is that the workforce becomes dominated by time servers who turn up, do as little as possible and then go home; for why bother? If, in trying to do the best they can under prevailing circumstances, they are met continuously with ferocious criticism, what is the point of labouring so hard? After all if CW's as a generality are branded as lazy and incompetent what have CW's got to lose by being such?

This almost totally describes the endemic attitude in care, both in the private sector *and* in some not-for-profit care providers too.

While I wish to make no excuses or condone wilfully poor care – which is never excusable and should never be condoned - the fact that CW's have to operate in an environment that is not just chaotic at the point of care delivery, but also totally dysfunctional at the management level, critically impairs their ability to do their job to any degree of acceptability.

However it is when Mr Ghani and Ms Lockhart turn to the direct causes of state failure that their observations become really illuminating if we see them through the prism of care.

Once more Mr Ghani and Ms Lockhart sum up the situation in care companies precisely when they go on to describe one the key factors that mark out failed states – the misalignment of objectives - as they note –

"The syndromes of a misaligned organisation are that the staff, resources, culture and processes are not geared to serving a common objective. Among the many useful ideas on alignment......one stands out: in order to ensure ongoing success, *a financial perspective alone is insufficient* [my italics].........

The same assets [staff, resources, culture and processes] combined in various ways can produce very different outcomes. Assets can either bring stakeholders together behind a common purpose or drive them apart. When a system is aligned, it acquires a synergistic relationship between actors and levels to achieve its goals. *When it is misaligned chaos, division, and confusion ensue* [my italics]."

The two points to take away from the above in terms of "the care industry" are first is the central conflict of interest that runs right through private for-profit care – that is between care of the best quality vs maximisation of profit; and second that in order to in any way resolve this conflict we are left with either an endless inbuilt ratcheting up of costs or a critical decline in care quality – either of these two unpalatable options become the only way to cut the Gordon knot we have tied around the throat of our social care sector.

Firstly then the original sin – from which all manner of baleful effects sprout – of the central conflict of interest in for-profit care.

This conflict of interest is so obvious - so glaring indeed - that it hardly needs expansion; however, for clarities sake I will do so; if the maximising of the quality of care means a company loses money then it will not long endure, therefore there is an explicit a trade-off required in providing care of lower quality but which can be delivered in order to turn a profit. Worse, in order to maximise profits care quality will have to decline further in order to meet this objective.

This is a classic misalignment – the very profit motive means that care quality suffers.

Not to labour a point let me put it another way, what for-profit care providers implicitly do is provide the best of care *for a price.*

The point then becomes – at what price?

The fact is that in care every "efficiency" gain that can be squeezed out of private provision in order to provide "better value for money" - the supposed attraction of introducing "the market" into any former public sector domain – comes at the cost of care quality due to the need to preserve margins. However, in care there is base level of care quality below which providers cannot sink – at least notionally – a point I shall return to later – therefore there is "price floor" under which costs can be cut no further without care totally failing. Therefore care costs are locked into a level that will have to be met in order that capacity from the private sector can be maintained – the alternative is providers exiting the market in droves can social care capacity sinking to dangerous levels (there is already a crunch in social care capacity and we have not yet to see the full effects of a bulge in the numbers of elderly and sick in need of long term care).

If there is a floor though, there is certainly no ceiling, there is no "market" orientated method of restraining costs, without, as noted above, putting either care quality or capacity in jeopardy.

However this is not the whole extent of the mess we have placed social care in, as more and more private providers have even encouraged into the care sector public sector capacity has been shrunk to a withered rusty nail in a crumbling wall, this mean that private providers of social care are rapidly becoming the only sources of capacity to absorb the rising need for social care brought on by an ageing population, in other words they are the piper who needs to be paid. As private providers have effective a strangle hold on the social care sector, in the future they will be able to use this dominant position to impose ever rising prices, this fact is almost an inevitable consequence of letting market forces loose without any mechanism to restrain long term costs. In brief we have traded the minor cost savings produced by "outsourcing" care for heavy, long term, open ended liabilities.

As David Freeman noted in the documentary on the failed, corrupt power company Enron, The Smartest Guys In The Room –

"A free market is goddamn expensive for the consumer."

It should be obvious from the above that there is a moral dimension to the two points raised - and that is how far are we willing to tolerate barely adequate care in order that costs may be kept down. Can we as a society tolerate the fact that our elderly, infirm or sick will receive only the most basic of care without a need for the state to raise additional revenue or cut services elsewhere in the social care system? Do we believe there is a limit to the quality of life that we can "afford" those needing social care? Or will we live in state where those with means will be able to buy "more" or "better" care and those without will have to do with the bare basics?

For this is where matters are leading – indeed we are already far down the path of personal provision for care – what we are witnessing is a massive transition from state funding of social care to the individualisation of costs - and the obvious outcome of this is that those with means will be able to afford a "more" and "better" care while the rest will be left with "less" and "worse".

At present LA funded care often only just covers the cost of "board and lodge" and families must pay a "top-up" fee. As rates of payment from LA's decline due to austerity and the increasing numbers of the elderly needing residential care kick in this "top-up" is also rising as is the number who are privately funded.

The term "privately funded" is itself a rather misleading in our present social context, summoning, as it does, images of the very wealthy with money to spare resident in palatial accommodation. The reality is that we are now living in a system of social care (and increasingly healthcare too) that is ever more notionally socialised but in fact is ever more paid for privately – that is some or all the costs are met by the individual – this is not the domain of the super rich; but is instead true in the generality of people in need of social care.

Some figures and a bit of economic theory needs to made here, however dull it may seem.

The ceiling by which all care must be "self-funded" – because of the conditions noted above – the rising cost of residential care and the falling level of LA funding – is catching more and more people. Currently those elderly with assets of £118,000, that includes house and savings, must pay for their own care.

Even a modest house and some small savings kick most elderly over this limit so that almost all of them currently have to pay for some or all of their care; full state provision only kicks in at £14,250 - so that means that only when the asset value of an elderly person sinks to this level will they receive assistance (NB it should be noted however that between these two figures some offsetting state aid will be provided financially).

The current government seeks to fix a cap on the total amount any individual will have to pay for their care at £72 000 but this figure and statement is at best ambiguous, and at worst wilfully misleading, it is also economically pie in the sky.

First the misleading part – although the £72 000 is strictly true is only applies to *nursing care costs*, the residential part of care – hotel costs as they are called – the price of bed and board effectively – would still have to be paid even after the £72 000 figure has been breached. In short the £72 000 is almost meaningless as most people would understand the notion of the word "cap"

Further – this cap is a promise that is bound to be broken as it unsustainable in the light of the ever rising costs for social care - the reasons for which we outlined above. It is a promise based on no realistic assessment of an ageing population and shrinking, narrowing tax base..

It also ignores the elephant in the room that we have already alluded to – and that is that it does nothing to address the stranglehold that private providers have over the costs of care – by handing the care of our elderly to private operators there is no way that costs can be tied down nor a "going rate" enforced – no more than government can introduce a cap on electricity costs in the long term. This is where the "god damn expensive to the customers" part of "a free market" applies; be it £72 000 cap or the £113 750 that currently applies, the lion's share of healthcare costs will be borne increasingly individually.

But money is only part of the care equation, a much more important question is what kind of care will be provided for the money paid regardless of who pays it.

In my time in care for the elderly and those with learning disabilities – as I have more than covered previously - I have seen more and more the unsuitability of the care offered because it places those with extreme needs together with those with less extreme ones. The confused and physically failing elderly are increasingly housed with those with severe antisocial behaviours – shouting, screaming individuals, physically violent, or physically, mentally and sexually threatening ones, faecally incontinent and challenging service users (i.e. those who in effect will throw or smear excrement over themselves or others in communal areas and sometimes invade others' personal rooms to do so), long-term and now aged mentally ill individuals, and those with no concept of personal property or space who compulsively enter other individuals' rooms and "expropriate" property - all are frequently put on the same units as those who are defenceless and vulnerable. Harried care workers, undertrained to deal with these multifarious issues, are forced to make good what they can the deficiencies of the facilities and organisation of the services that they work in and sometimes are consumed with conflict management over providing actually care and support.

Two examples will illustrate what I mean.

In one home I worked in EMI nursing patients were mixed with residential patients - this lead directly to a conflict in care needs between those who had a high degree of personal functioning with those with very extreme behavioural issues and can be described – brutally, but I believe frankly – as placing lions with antelopes with only a couple of keepers to keep the two separate (the zoo metaphor here is no way meant to demean service users and I utilise it only to reinforce the "caged" nature in which we place our elderly, sick and vulnerable).

One resident I supported was particularly susceptible to the effects of the challenging behaviours of others – diagnosed with latent paranoia that could rapidly become acute depending on environmental conditions, it was known on admission that they jealous guarded their own living space and felt existentially threatened if this was invaded, were easily and seriously affected by unpredictable behaviours in others and liked to keep themselves and the things they possessed – either temporarily – plates they were eating off, cups they were drinking out of - or permanently – their own personal property – scrupulously clean and unmolested. Despite these mental health issues they themselves presented no challenging behaviours and were mostly placid and communicative on a one-to-one basis, although were deeply reticent about wider sociability. They were placed on a mixed EMI residential floor where there were several wanderers who would freely enter other service users rooms, use their facilities – like the toilet or bed – and would habitually take things not their own, as well as individuals with personal care issues and – as noted above – a propensity to manifest behaviours by the spreading faeces or urinating in inappropriate places.

This resident on arriving was rendered into a state of horror akin to shellshock, within the first week they had observed a service user deposit a lump of faeces on a dinner plate in the dining room, had another resident eat from their plate, and had had several items of property taken from their room never to be seen again. Unsurprisingly they swiftly retreated to their room and within the first week of them being in residential care they confined themselves to their room on a 24 hour, 7 days a week basis, locking the door and becoming suspicious and hostile to any CW who entered the room or who tried to support them, and occasionally becoming very challenging if they felt like they were being coerced to leave their room. This withdrawal meant they ate all their meals in their room, slept for large portion of the day, refused to shower or have their bed changed and presented with all the symptoms of paranoid depression.

One nurse tried with exceptional patience and understanding to encourage socialisation - fearing a total withdrawal could become permanent unless the issue was tackled quickly – an eminently sensible and laudable aim. They had some success over the next few weeks until their efforts were summarily stopped following an incident when the service user in question became challenging and withdrew because - again – other residents ate off their plate – the nurse then found themselves blamed for the incident and disciplined for being "overly coercive" in their encouragement of the resident to socialise and for triggering another paranoid episode.

Indeed paranoid depression is what the nursing staff, backed up by the visiting psychiatric consultant diagnosed; yet, bizarrely and lamentably, they put this down to the underlying condition that they had originally diagnosed and that had brought this service user into residential care. No mention was made about the unsuitability of the placement or the fact that their downward progression had been, at least to some degree, and, to my observation, for the most part, *due to other service users behavioural issues.*

In this instance more medication was brought in – one of the classic types of the misuse of medication that had been discussed previously.

This medication, although admittedly reducing the paranoid symptoms – the service user now allowed CW's to bathe them and change their sheets once a week – also meant they slept for longer and longer periods with the result of them leaving meals because they were simply too exhausted (doped up perhaps would be a better term) to eat them. I myself noted putting food before this service user and coming back 5 minutes later only to find them slumped over it. I attempted to rouse them, which was extremely difficult, and when I succeeded I tried to encourage them to have something to eat and drink, at this they became irritable and then defensive. Knowing they had issues around their personal space and fearing tipping them over into another paranoid episode (and also more than wary of meeting the same fate as the above mentioned nurse) I had to leave them. The upshot of this was they began to rapidly lose weight and suffer dehydration – a situation that was barely managed (i.e. the onus was placed on CW's to see that they were adequately fed and watered in the face of medication and environmental issues that forestalled these attempts at every turn).

A crippling irony in this case was that there was a suitable place within the facility to which this service user could have been transferred – that being the general nursing floor. This floor was quieter, had no challenging residents and no wanderers that could have caused problems. However they were not transferred here for two reasons – Firstly their LA would not cover the full cost of the more expensive "nursing" bed and their family were without the means to provide the "top-up" fee to meet the shortfall - which was quite substantial. Secondly the nursing floor rooms were being reserved for more lucrative cases that were potentially in the pipeline. The result was that this resident lived out the last months of their lives unhappy, unsettled and surrounded by the very people they would have run from if they had only had the freedom to do so.

The second example also involves the nursing floor. A resident was admitted with issues well known in their admission assessment; although they had no real challenging behaviour issues they were habitual night wanderers, were invasive of others rooms and "acquisitive" of others property (taking things and hiding them, sometimes permanently – i.e. losing them) and had a propensity to try to "help" other service users which included pulling other service users out of wheelchairs or seats, "arranging" beds that could cause serious issues regarding safety and potential to harm – placing pillows over residents faces or bedding over their heads – as well be being a danger to themselves due to their risk of falls.

Despite knowing about all these behaviours the service user was placed on the general nursing unit because there were no places on the EMI unit – in other words they were admitted simply to provide revenue even though it was known by the home manager and more senior executives in the company that this placement was totally unsuitable. The result was this service user invaded the space of, among the other service users, two who were receiving palliative care because they were in the end of life stage and another resident who was paralysed from the neck down and therefore was powerless to prevent the new service user walking off with their possessions. Objections were raised repeatedly by the care and nursing staff but this resident remained on the nursing unit for two more weeks before a vacancy opened up on the EMI unit and they were eventually transferred; though not before the two service users who were basically dying had had their last hours (and their families – sharing in these last precious hours with them) disrupted by constant intrusions.

Where this left dignity, respect and care I leave it to the reader to make their own minds up, but this leads us on neatly to how dignity, respect and care are treated in care homes as opposed to profit. (The final irony in all this was that while these issues were taking place it was "Dignity Awareness Month" in the home in question, where staff were encouraged to pay £1 for a "Dignity Ribbon" to display their "awareness" of the issue – strangely I never saw any management or company executives divvying up their quid for the dog and pony show that this was, and oddly I never found where that money was going to.)

Both these examples clearly demonstrate what is common practice in for-profit companies – care is dictated by cost and revenue.

The latest buzz phrase doing the rounds in care is Personal Centred Planning/Care – this is supposed to put the service user front and centre of all care and to tailor that care to their requirements and needs, ensuring they are treated as individuals and that they remain as independent and autonomous as possible given their medical and psychological conditions. All very well and good, no one wants to be treated as an anonymous cog in a bigger machine, no one wants to have their preferences rode roughshod over because they simply do not fit in with the wider organisation of care and no one wants to have their independence sacrificed to expediency and a one size fits all care home regime, no matter what their mental state, no matter what their physical health, no matter if they may be able to communicate their wishes or not.

Yet this ignores a central contradiction in residential care – how can you house the elderly and infirm cheek by jowl with each other – many with deep and antisocial behaviours, many with complex care needs, all with their own histories, lives and family demands - and still retain anything resembling Person Centred Care. The short answer is you can't, or at least you cannot with the typical staffing ratios found in care.

In the average care home a ratio of approximately one care workers to 8 residents exists (this is discounting nurses who, in care homes as in the NHS – as the Mid-Stafford Hospital inquiry showed - actually deliver little "hands on" care but are consumed by other "higher" considerations than actual physical patient care) rising to 1:10 or 1:15 at night. What may shock any reader is that there are no minimum staffing levels that can be enforced by regulation – there are only guidelines. Let's go through that again with the appropriate emphasis - _**there are no minimum staffing levels that can be enforced by regulation – there are only guidelines**_

(NB - It should be noted here though that a care home _can_ fail a CQC inspection if the inspectors _feel_ that staffing ratios are inadequate, the very fuzziness of this concept - based on perceptions and feelings - and added to the potential for care home managers and owners to finesse any short staffing - makes this supposed check on low staffing almost useless in practice.)

That bald fact – the lack of enforceable regulation - should stun anyone who thinks that the elderly are well catered for in the private care home sector. What should shock them even more is that **_these ratios frequently are not met_**. I have worked a shift where there have been only 2 CW's (when there should have been 4) and one nurse to cover over 30 residents on 3 different floors, I have known a day shift when there have been only 4 CW's (when there should have been 7) and 1 nurse (when there should have been 2) to cover 40 residents on 3 different floors; further I have known 1 CW (when there should have been 3) have to cover 15 residents during the day on an EMI unit– three of whom were on 15 minute observations because of violent/sexually predatory behaviours, 4 of whom need assistance with feeding and hydration and 8 of whom were doubly incontinent and required two CW's for personal care (that is they needed frequent changing by 2 CW's to avoid them walking around soiled or soaking wet).

Even without staffing guidelines it doesn't take a genius to work out that this not only is well past unacceptable but highly dangerous. What was this single CW on the EMI unit during the day to do – I ask – when – not if – any service user needed changing? And even if there were 2 CW's on the unit what would happen during personal care for any service user while both were occupied? – leave all the others unattended? Including the 3 potentially dangerous (toward other residents) service users? Or do should they just leave service users wet and soiled? How, I would further ask, would this single CW check that all the service users had received enough to eat and drink? How would this single CW get their required breaks? And how, finally, does this square with Person Centred Care?

It may be argued that here I am taking extreme examples to prove a point, and, to an extent, this is true; however it was also far from uncommon - in the case of only 4 CA's having to cover 40 residents on three different floors this happened at least once every 2 or 3 weeks in one of the homes I worked in.

Even more common than this was the situation where only 2 CW's were left supporting 15 EMI service users – this occurred at least 4 times a week…. habitually. While this is preferable to 1 CA *and which – with 1 nurse – met suggested guidelines* it was impossible to offer any kind of quality *and safe care* in this environment. Yes there may have been enough "bodies", according to guidelines, to cover the floor, but in fact this staffing was inadequate considering 3 service users were on 15 minute observations for challenging behaviour.

How are these two CW's supposed to bath or shower anyone when this would effectively take them out of being able to monitor those under observation? Was a single CW supposed to bath or shower a service user alone – a practice fraught with danger – think an elderly person with poor mobility, think slippery floors, think wet soapy body? Then what about those service users for whom it took 2 care staff to move – do they simply not get showered or bathed? And when are either of these two CW's to take break? Finally, and once more, where does this leave Person Centred Care?

An example here will show that poor staffing ratios are not just an aberration but the norm.

A unit in a facility in which I worked had 2 CW's to 15 residents – a ratio just inside the suggested ratio of 1:8. All well and good - only 6 of these service users needed hoisting whenever they were to be moved from one place - say bed – to another – say a wheelchair; four more residents needed total support with all their food and drink intake and 10 were doubly incontinent. So essentially at mealtimes 6 people needed hoisting and moving to the dining room, 4 needed to be fed and 10 needed regularly changing – what this meant in practice was that the 2 CW's on the floor were constantly busy merely addressing the immediate needs of all the residents, and even then it was not always possible to give the full time to those who needed feeding – among whom was one that it was not uncommon for them to take over an hour to consume only half the food on offer.

The result was almost inevitable – half of the residents started losing weight and, more than this, at least three quarters of the residents showed classic signs of dehydration – very dark urine, confusion above that which they normally presented with, constipation and drying skin.

This, in my view, was not the fault of the CW's – they did their level best to meet the needs of all the residents - but it was simply not possible to meet all of their needs all of the time. Staff complained over and over that they simply did not have the resources to adequately care for the service users on that unit and pointed to the weight losses as evidence that they could not spend the appropriate time with vulnerable residents. These pleas fell on deaf ears *because the staffing ratio met the guidelines regardless of the manifest facts that care was failing.*

The truth is that in all private care homes staffing ratios amount to what I have alluded to above – having the requisite number of "bodies" on a unit so that it does not appear as if staffing has fallen below a certain ratio. Whether these bodies can actually deliver any quality care – or even the bare minimum of care - which I would argue keeping service users safe, well fed and hydrated and bathed or showered regularly consists of - is beside the point, the only fact that care home operators are interested in is that the numbers add up on paper and to maintain appearances.

Further to this most care home operators take no account of not just the raw ratio of care staff to residents but the nature and needs of those residents. A ratio of 1:8 may be more than adequate if all the service users are mobile, can feed and hydrate themselves and are continent; this picture radically changes though if we say that 5 have no mobility and are doubly incontinent, 3 more are mobile but are also incontinent and 3 further ones have extreme challenging behaviours – in this equation 2 care staff are hopelessly outgunned by the needs of the service users and the demands of the service.

This very fact – the needs of the service users as well as their sheer numbers – is supposed to covered by the dependency ratio; this is a kind of formula that accounts for the care time required by service users according to their needs, and examples can be found at

http://www.rqia.org.uk/cms resources/Staffing%20 Guidelines%20for%20Nursing%20Home%20Version.pdf pages 4 and 5.

This dependency calculator is then done for each service user and total care time added up and from this comes the number of care staff that will be required to staff the service (page 6). This scientific measure is all very clever but I would urge the reader to look and see where it actually allots time for Person Centred Care – that is not just a raw calculation of who needs however much care but *how that care is delivered.* With such an antiseptic and clinical calculation (or "tool" as they are now called – if you hear that word fear and tremble, it means care has been reduced to formulae and tables) what is being done is tacitly militating against *what is supposed to be the guiding standard of care* – its personal basis. The calculations shown reduce each service user to a set of needs and allots them a minimum amount of time to address those needs, there is nowhere in that calculation that says that Mrs X likes to go to the toilet just before dinner and needs to be hoisted on and off her convenience, nor is there mention that Mr A should never be left out of line of sight at any time night or day.

By the reduction of individuals to raw needs – care reduced to a time and motion study – it is almost forcing care staff to view each resident as a fixed number of tasks.

In the immediate above I mentioned about a notional Mr A who "should never be left out of line of sight at any time night or day" and this leads us on to a gaping hole in care that is almost never addressed, regarded or fixed upon as a systemic problem that is almost inviting catastrophe in every care home in the country – and that is the ratio of staff to service users on a night.

In the above quoted document the table on page 2 shows a staffing level of 1:5 (a fanciful ratio to begin with – I have hardly ever seen this high a ratio in private care homes) falling to 1:10 at night. This is typical and understandable to anyone *outside the care field* but incomprehensible to anyone who has ever worked nights in a nursing home. The common perception is that on a night all service users will be nicely tucked up in bed and sleeping soundly, so fewer care staff will be needed.

Fine.

The problem is when you dig a little deeper. Night shift typically starts at 8pm – does anyone you know think it normal or desirable to be in bed *before 8pm*? – I certainly cannot think of anyone like that. More, do you ever have to get up and go to the toilet during the night? Great, if you can walk there, but what happens if you need hoisting out of bed into a wheel chair, along to you toilet, be hoisted again onto the toilet then it all done in reverse once you have done your business – a process that is supposed to require 2 CW's to be achieved safely. Yet the guidelines say that only one care staff member is needed on a night so then, I would ask – how is that one staff member supposed to support such a resident when a hoisting procedure demands two CW's to achieve safely?

This is not a theoretical scenario – it is one I encountered nearly every night in every care home I worked in and leaves a CW both impotent in the face of a residents needs and totally unable to meet the most basic of them with regard to dignity and respect as well as the supposed golden rule of individuality – the ability of a service user, say, to toilet themselves if they so wish – no matter if they need intensive support to do so. After all, this what care homes are supposed to do – meet the intensive needs of service users that can not be met in their own home - yet if low staffing prevents CW's from meeting these intensive needs, what then is the point of care homes and residential care?

In this situation, lacking in basic adequate manpower, what actually happens is that you as a CW have to tell an individual (yes - *an individual* – that basic tenet of Person Centred Care – that slipped your mind? No matter it seems not to have crossed most care home operators minds when it comes to night staffing ratios) who has most probably proudly always taken care of their appearance and personal care habits all their long lives, that they have to piss in their pad – basically wet themselves – or else try to get them on a deeply uncomfortable rigid plastic bedpan while you watch over them to see that they don't slide off it or spill the contents all over their bed. How humiliating is that? How depressing for any CW with any notion of care within them. And yet as a CW – whether you care or not - you have no choice. To try to hoist a resident on your own is to put them at a gross and unfair risk – after all if something did go wrong and you injured a service user, or worse, no one would thank you for putting Person Centred Care first. Instead to have to go against the grain of all your training and instincts and see someone slowly degrade themselves simply because you have not the resources to support them.

This leads us neatly to the actual continence pads themselves - which are supposed to provide a modicum of decency to incontinence by trapping urine and faeces and allowing care to be undertaken in the least invasive way possible.

The problem is that most local authorities require an evidential record that an individual service user is in fact incontinent. To get this a continence nurse needs to visit to assess a service user and usually will require a record that needs to be kept of every episode of incontinence *without a continence aid*. That means – to put it brutally – an individual has to urinate and soil themselves on a regular basis before they can gain access to the very continence needs that are manifestly required.

Should they get them their problems are hardly at an end, pads are severely rationed per service user, each has an allocation of four – yes that's 4 - in any 24 hour period.

Think this is adequate? Then perhaps you would like to undergo the same indelicate test that service users are forced into – see how many times you have to go to toilet in any 24 hour period, then see if you "go" less or more than 4 times in any 24 hour timeframe (urinating and defecting yourself are optional here, for the incontinent they are not).

Any more times than 4? Tough, you've used your allocation – looks like you'll have to sit in your own urine and/or faeces for whatever remains of the 24 hour period until you are *entitled* to go again with any degree of security that you won't be getting your clothes or your bed drenched and/or soiled.

Even this experiment does not do justice to the fix that the incontinent and in care are in, most especially the elderly who often, due to age, ailments or medications they are on, experience particularly heavy flows of urine; also, as mentioned in a previous chapter, they are often pumped full of laxatives, thus they urinate more and also have more frequent and extensive bowel movements. I have found it far from uncommon that I have had to change service users 3, 4 or 5 times in *a 12 hour shift period* because they are wet or soiled and uncomfortable. This means that they run through their allocation of pads in no time leaving a shortfall; as no more will be forthcoming until the next order is due, if continence aids are to be provided once more the family have to pick up the tab.

How's that for a choice? Pay up or your relative is going to be enjoying a wet and dirty next few weeks.

In fact that doesn't happen for obvious reasons – any care home letting residents soak and soil themselves because they have run short continence aids would find itself in trouble in short order. Instead the service provider – care home owners - have to provide continence aids and bill the families and hope they pay up. Because no service provider wants to be stiffed by families refusing to pay up care homes guard pad allocations as if they were woven from golden thread.

The rationing of pads is such a serious issue in care for the elderly that in one care home I worked in the manager had them under lock and key and they were handed out to CW's only on request – that's right, when a resident was wet or soiled a CW had to go to the manager to ask for a fresh pad. On night shift the care team were handed their "ration" of pads for the night.

A horrible irony is that many of the elderly in our care homes today will have already experienced rationing – during the war. Now, when they notionally live in a land of relative plenty, in one of the richest nations on earth and with the 6th largest economy in the world, and yet they have to endure the fact that they are "allowed" only so many pads per day.

Of course CW's are endlessly resourceful in finding workarounds to these strictures – they often will use the pads from a less incontinent individual for one who is more so. However this is serious offence – it can be classed as theft of property, as a pad allocation is effectively the "property" of a particular service user. This can be classed as a criminal offence with the attendant penalties associated with that crime – that is the police, arrest and prosecution. At the very least using one service users' pad on another is a serious disciplinary infraction that can lead to instant dismissal on grounds of gross misconduct.

Even when continence aids *are* provided there is no guarantee that they will be suitable ones. Time and again I have found that service users have been assessed for pads that deal with only mild and occasional incontinence when in fact they are totally and frequently doubly incontinent; in such situations as these (and they are the rule rather than the exception) continence aids are next to useless in keeping service users relatively dry and/or clean.

Further, unsuitable continence aids set a vicious circle in train that defeats their purpose – the less suitable they are the more often they will "fail" i.e. – let a service user become wet through to the clothes or bed sheet – the more often they fail the more often service users will need changing and the more fresh continence aids will need to be used after changing – inadequate and unsuitable continence aids are therefore used more quickly than more suitable ones. The fact that this is false economy does not appear to have occurred to continence nurses, LA's or the NHS – the only point that counts is that they are cheaper than better suited continence aids and that's that; as long as costs are kept down that is the limit of their interest, the fact that the consequences of this are pushed back onto CW's - who have to deal with the effects - and service users - who have to live with them - are of no consequence, after all, the bean counters do not have to wear them.

Finally, in my experience, service users are rarely, if ever, reassessed as their condition deteriorates, to see if their needs as regards continence aids have changed. Often in the late stages of dementia service users' needs as regards continence aids increase, yet often their assessment does not, leaving CA's once again having to try to make a slowly dying individual comfortable in their last weeks with continence pads that are wholly inadequate for their altered needs.

Continence issues for night staff – when meeting toileting needs are most difficult – as noted above – pale next to the problems raised by the management of service users who simply do not sleep during the night.

One nailed on guarantee of working with dementia sufferers is that they have chaotic sleep patterns. This means they often fail to sleep on a night, and, if mobile, will wander freely. Most commonly this practice will take place on an EMI floor where residents are suffering from a concatenation and combination of mental disturbances that make adequate monitoring of the security of individuals difficult to achieve – a wanderer will see nothing wrong in entering whichever room they like, and those who occupy the such rooms will not be able to understand their right and ability to lock their doors; therefore the only guarantee of their sense of safety and security - and their right to expect the integrity of their own private space - falls back on CW's.

Again we come back to dignity and respect as well as Person Centred Care.

A service user has every right – as much right as anyone living in their own home – to expect that their personal space will be protected from interlopers however harmless they may be, however unable they are to fathom the concept of the privacy and individual security; and yet this duty is often taken incredibly lightly even by very senior and very experienced care staff and clinicians as well as managers and executives. One of the most bone headed sentiments I have ever heard voiced to a relative who complained about these breeches of privacy - and this by a senior, vastly experienced and otherwise very efficient and competent nurse - was that room intrusion is only "to be expected on an EMI floor".

Wrong. Wrong. Wrong.

No one should have to expect that their living space will be turned into a public thoroughfare, this is a fundamental and basic right and should be front and centre of care, after all if we as professional carers and clinicians cannot ensure the basic integrity of space then what the else are we there for?

On a night though the very securing of this space becomes untenable, again simply because the manpower is not there to assure it. A strafing ratio of 1:10 or 1:15 (and 1:20 is not unusual either) means that there is no way a night carer can keep eyes on everyone all the time. One of the most difficult issues on an EMI floor at night is keeping tabs on any wanderers while still attempting to provide care for the rest. How can you deliver personal care and still keep tabs on someone who can slip into any room at any time? The truth is you cannot, therefore your old and confused mother, or father or brother or sister, aunt or uncle, may find strangers walking almost at will into their bedrooms at all hours of a night – a horrific betrayal of – as I mentioned above – one of the most basic rights that we should all expect as long as we are breathing air. This issue becomes even more vexed if we move to the next bugbear that in my time in care depressed and irritated me almost to distraction – mixed sex units.

It always stuns me that we simply do not expect our old to have a choice about who they share a living space with, further, that the care system in the UK tolerates the fact that that our elderly "will not mind" or worse "will not be of the mind" to notice that after 60 or 70 years living either largely alone or closely with another person to whom they are devoted or tied to, surrounded by intimate relations and dependent on the loving family unit, that they should be, in their last - and therefore their most precious - years, are thrown into sharing their lives with other strange men or women.

Imagine yourself not only turfed out of the only home you may have known your entire adult life and which you have shared with someone you have spent years in the company of and who you have loved dearly and suddenly landed on a locked unit with people of vastly differing social classes, backgrounds and, ultimately, sexes. How would you react to this environment? Perhaps it would be ok if you were of the gregarious type who always enjoyed an enlarged and lively company, hard luck if you treasured your personal space, silence and autonomy. Even harder luck if you have lived and loved only one other person all your adult life and now have the attentions and even depredations of other males *or females.*

This last point is important – we often associate the unacceptability – if we consider it at all, which, largely, and to our shame, we do not - of mixed sex units as being because females are placed together with males and so the chances of inappropriate behaviour by such males, or even outright abuse by them, are heightened. While in the main this is the most pressing issue, it should also be noted that males – who because of their shorter average lifespan compared to females often find themselves in the small minority in care homes – may also find themselves the subjects of unwanted and inappropriate attentions of strange women. More than you would expect I have seen females attach themselves to males either through misidentification of them as their (long dead) spouses or through the natural effects of affection.

I can name numerous examples where males have been deeply upset and have sought refuge from females who are effectively hounding them with unwanted attention. Once more I would call on you to imagine your father or brother, who after a lifetime of fidelity to one women or man now finds others trying to put themselves in the place of that person – how uncomfortable and anxious would that make them? Most especially if they believe that their spouse is still alive. Yet this fact is frequently unrecognised and instead a lazy assumption is made that males will "enjoy" or "welcome" such attention or that they, as I have heard it said, *"must feel like a kid in a candy store".* Such stupidity is sadly rife.

However having said that the most problematic issues do arise on the side of vulnerable females and predatory males.

It is a fact that even with the best of monitoring, even in the best of care homes *the protection of vulnerable females will be impossible to guarantee* as while the most obvious predatory actions may be prevented it is difficult; no, check that – impossible - always to be aware of the less overt signs of depredation. Further than this though – and this is an important point to make in respect of the mental anguish that can function below the surface – actual outright abusive actions need not happen for abuse to take place – there only needs to be the implicit threat; in other words *a vulnerable female needs only to feel threatened for abuse to happen* – I reemphasise – __the threat of abuse perceived is enough for actual abuse to take place.__

That may be a difficult concept to grasp in the abstract so let us put an example that we can relate to – if you *feel* threatened by another person the actual threat does not necessarily need to be followed through – you only need to believe that at some time or another *it will be followed through* to amply be threatened and victimised. To put more baldly – if someone displays an implicit or explicit threat to do you harm it doesn't matter that the other party does not intend in their own mind to carry it out, they only need to implant in *your* mind that they will, for it to have weight – a man does not need to actually rape a woman to abuse her, she only needs to believe that he at some point will. We, in society, would surely condemn and hope to see prosecuted or otherwise sanctioned a man who held the threat of rape over a terrified woman to have control over her – and yet we put with this type of abuse every day and night in care homes.

Lest it be thought I have drifted into hyperbole here a couple of quite extreme examples will suffice – and by extreme I mean the actions were extreme, not the circumstances – as I encountered the same circumstances over and over again, all the time I worked in care.

On one unit in which I regularly worked there were two male residents who exhibited predatory and sexually motivated actions aimed at various women residents on the same unit. At any opportunity they would grab females on the bottom, breasts or in the genital area, force their tongues down the throats of any woman they could get to grips with and push the hand of any female in the direction, or actually onto, their genital area. Less extreme were the actions they partook in nearly all the time – taking a woman's hand, invading her space, stroking them. Although these two men were under 15 minute observations it was virtually impossible to police the latter manifestations of their behaviour – the hand or hair stroking and the invading of personal space.

It will be argued that these actions were in themselves harmless - and I would heartily agree - if they were not in themselves precursors to the more extreme actions mentioned – in other words they were antecedent - and strongly suggestive - of more abusive actions to follow, therefore to prevent escalation these actions need to be stopped at the first sign.

This should stand to reason – an individual – even one very confused and mentally ill – often will follow the logical course that if one action is permitted then a following, more elevated, action will also be permitted – if the stroking of a hand is ok, so too will be the forcing of an invasive kiss.

There is also the other side of this equation to weigh – we as carers, or even just casual observers, may see nothing wrong with these actions, but how does the woman who is the object of these actions react? Sometimes the reaction is clear-cut – shouting, striking, and, if they are mobile and able, moving away - but other times the signs were much more subtle.

I once entered the lounge of the unit to find one of these predatory males sat extremely closely – virtually on the lap - of a very confused female, he was stroking her hand and whispering in her ear; the female was mobile but remained in her seat and her hand in the males hand – what harm you may ask?

The key was in her expression – stuck somewhere between extreme discomfort and distress, her face turned away from the male, her shoulders shrunk together in a defensive posture and her body as far away as she could get while still sitting in her seat. If this female was enjoying the attention then I would hate to have seen her with an absence of joy. This happened on a night shift when I was the only carer on the floor so I could not keep tabs on the two male residents together, so I did what I could by removing the male to his room to deescalate the situation and where, at least, I could observe either his door or the corridor outside it. Further, by removing the male I would have time to react and follow if I had observed he had left the room while my attention was elsewhere.

Had any abuse taken place here? Not anything I could write up; however, and referring to what I said above, was this not enough? Wasn't the female involved entitled to sit in a communal area without her personal space being invaded, without intimate actions being undertaken without her approval or consent, without *feeling* threatened? Of course I could not say *how* the female was feeling but I could take a damn good guess.

Although the safeguarding of residents was something I took very seriously on its own merits – that I firmly believed in the rights of females *or males* not to be molested or touched inappropriately – there was also the professional cost - if a serious incident had taken place the consequences would have rolled back on me. If an actual and overtly harmful event had occurred I would have been answerable. In such a situation, and its aftermath, the first question any relative would ask was - where was the carer? Even with all the paperwork filled in, even with every effort taken within my capacities and capabilities to safeguard the resident, my head would still have been on the block.

This raises a dilemma that shoots right through care and affects most seriously CW's who work alone, it deserves attention here both because the context naturally leads to it but also because it is worthy of raising on its own even at the expense of slight digression.

In this instance if I had witnesses a serious sexual assault would I have reported it knowing I would have been answerable for the result?

I would like to think yes.

However the question would remain.

[N.B.This is a critical point that needs some weighing – all healthcare workers, not just CW's, are often confronted, sooner or later, with an "event" that may not have been wholly of their making but for which sole responsibility rolls back on. The overwhelming urge in this situation is to cover things up.

The Francis Report into Stafford Hospital is riddled with such instances – care or clinical staff, confronted by serous incidents, rather than report them and risk blame, disciplinary measures, firing and even criminal or civil prosecution – took the Watergate route of burying the evidence instead.

While totally inexcusable, it can be understandable. To illustrate, let's look at the real life example noted above that I experienced; with no other CW to vouch for me and all the measures I may have put in place to prevent abuse and safeguard vulnerable residents I was horribly exposed. To show how much exposure I was open to let us suppose that something of the nature of a sexual assault had occurred - and I had reported it - any investigation would sooner or later have asked me if I considered the situation on the unit where I worked – with two predatory males and no hope of keeping tabs on them – was safe, if had said yes they would then have asked how was it that I could not safeguard the well-being of vulnerable residents? If I had said no then they would have asked me why I had not taken measures to address the unsafe nature of the situation, by reporting it to senior management or to the local safeguarding board – in the absence of doing either I would have been in deep manure.

In the light of this example it is therefore not surprising that healthcare workers, faced with such a lose-lose choice find it better to cover up.

The duty of candour introduced in the wake of the Francis Report – which places a duty on healthcare workers, and organisations, to admit to failures in care - will do nothing to address the dilemma – the fact remains that reported incidents of whatever nature are the tip of the iceberg of serious care failures, for every one that reaches the public domain there are dozens more than remain hidden, covered up, buried.]

The second example I raise in regard to safeguarding vulnerable service users poses serious questions to all the care staff about the rights and freedoms of all individuals in a home.

One service user I supported masturbated almost constantly day or night and made lewd comments to any female in his ambit. This often took place in communal areas. The suggestion was that we remove the gentleman to his room whenever he began to masturbate. However, as he masturbated almost constantly he eventually became wholly restricted to his room.

The question here surrounded deprivation of liberty. While the masturbation is socially unacceptable, and public masturbation would be criminal in most circumstances, how, still, were this gentleman's rights - as far as being able to socialise - inhibited. Most observers would view this as a no-brainer, a straight up and down choice for the gentleman in question – if he wanted to socialise then he would have to refrain from masturbating, failing to restrain himself meant his actions were in themselves anti-social so he forfeited the right to socialise.

This choice though suggests a rational mind weighing alternatives and costs and benefits, a capacity that he lacked – why else would he be in a home? It also failed to take into account the fact that the masturbation was to an extent compulsive – was this therefore a manifestation of his dementia? If so this posed a dangerous precedent – if one person could be isolated because of one of the manifestations of their disease what of others' anti-social habits? What about compulsive spitting, shouting or smearing of faeces – would these too be sanctioned likewise? The sexual nature of this anti-social habit were the sticking point, literally, in this context, however the comparison still stands. And it is a lingering issue in care. While the mind may fail, the sexual appetite remains, sometimes, as a direct consequence of the disease, it is actually heightened. All males that I encountered in residential care masturbated to a certain point, and many women too.

Both these examples point to the fact that should be manifest to anyone who ever set foot in a care home is that residential care of this kind is almost universally inappropriate and totally contradictory to the professed aims of Person Centred Care. In fact it can be said, without too much exaggeration, that it is in fact inimical to the dignity, respect and liberty of our aged, infirm and ill, it shuts them away in what amounts to, as I have already said – prisons with better wallpaper.

So what exactly does Person Centred Care amount to in care homes? And if it cannot be delivered what is its use? The short answer is that Person Centred Care is impossible in care homes as they are constructed (both in the physical building itself and as institutions) and as such it continued parroting is, I would respectfully submit, a load of tosh.

A more extended answer is that it is a chimera set up to give the *impression* of care, as if what we do is something more than battery farming for the demented where quiescence rather than fecundity is the object. What is aimed at is not care itself, tailored to the individual, responsive and flexible - *but the illusion of responsive individual and flexible care*, it is meant to dazzle the eye of the gullible, confound the senses of the naïve and distract attention from what is hiding in plain sight – a flawed system that satisfies no one except a remote government, desperate and exhausted relatives who have reached the end of their resources in looking after declining relatives, and care companies who hoover up public funds knowing that the services they provide fail on almost every count of compassion and dignity. It is in some respects a mirror of what modern care homes are – on the surface gleaming, smart, hospitably decorated and clean - yet the façade hides loose wiring, dodgy plumbing, and fittings that fall off after a couple of months - so too with Person Centred Care – it looks good on paper but in practice falls apart due to the contradictions in providing any personal touch in a multi-occupancy facility.

It's a sham and betrayal.

A last point to be made here is that Person Centred Care is supposed to be administered mainly by care staff, yet they are under resourced, in both staffing and time. Barely able to provide the minimum of care and safety they have little or no time to consider the individual in all their variety and vicissitudes. It is an irony that always struck me in one care home I worked in that part of the "activities" that were organised were "One-To-One sessions" – time when the full attention of a carer was devoted to one resident – for 15 minutes. It is not just the brevity of this close personal interest, it is the very fact that what should be basic care – spending time to pass the time of day with each resident cared for - was somehow classed as a "bonus" - something extra, a luxury *so munificently* provided. Most carers I know would love to spend more time with each resident but they simply cannot do so without compromising the basic care of others. Do you spend 10 minute chatting idly to a service user knowing another service user has not been fed yet?

Someday, a more enlightened day, in the future people will look back on care homes, with residents in various stages of mental and physical distress, rammed cheek-by-jowl with each other regardless of gender, level of needs and individual preference, trapped behind locked doors and shut away by the tacit consent of society and run for the profit of a few individuals and companies - and regard all of us working in the care system, and the mass public who tolerated such barbaric practices as – if not inherently evil - then at least as callous, indifferent and as stupid as those who smiled on homes for "fallen women" workhouses and child labour of ages past.

As I commented more than once to my fellow workers in my more frustrated and candid moments – in 50 years time those of us still alive are in danger of being put against a wall and shot for participating in such a system. For in a way we care workers and assistants are as culpable as care home operators – after all what's the difference – we both get paid for taking an active part in a broken and inherently corrupt system.

The analogy that always sprang to my mind – and I believe it still hold true - is that the relationship is rather like prostitution; care companies are the pimps, Care Workers are the whores and the service users are the punters. As my time in care went on I saw myself more and more as selling myself – and my care – for money and providing a "service" based on a minimum of social utility and a maximum of generating profit for those who directed me – i.e. the companies I worked for.

Finally my work almost become a valediction for all that is wrong in care in the UK - in a future enlightened time I hope people will wonder how we managed to so totally screw up the care of the most vulnerable and sick members of our society just when their needs rose to become of the highest priority.

It would be a damning indictment, but a true one.

However there is even more of a difficulty inherent in care in the UK than its rampant commercialisation and commodification of service users - it is in regard to the workforce and constitution of care provision - specifically the cultural and racial mix of the labour force. In almost every care service in which I worked the labour force was between a third and a half (and occasionally tipping over 50%) composed of foreign born nationals, most of them of a different ethnicity from the rest of the labour force *and* of the residents. This issue is one of the most explosive in care and has caused me much raising of eyebrows - and occasional hair - yet it is a subject mostly steered clear of by anyone talking about care - which I believe borne of sheer cowardice. No serious discussion or focus on care can be complete without taking into consideration the massive racial issues that often predominate and occasionally dominate care itself.

CHAPTER XIX

Race

No one wants to talk about race in care, more than this, subtly and implicitly – no one thinks you *should* talk about race in care; most people either believe that there is no problem with culture clashes in the care environment, or they *do* know there are problems but choose to ignore them, still others know and are familiar with the problems posed by race and racism in care but are too afraid to raise the issue lest they be labelled racists – so the problem remains hidden from broader public view because of an overriding condition that all comment about race is *verboten*.

That fact alone should set alarm bells ringing; anything that forbids discussion before it starts because of the inherent difficulties in it poses a grave threat to the coherence and quality in any social or economic field – dysfunction is allowed to flourish unhindered and even when it is exposed the true causes of it are simply skirted around.

The crowning irony is that care itself reproduces in microcosm all the major fault lines between races, nationalities and cultures in our broader society and is therefore a crucible regarding questions of immigration, multiculturalism, integration and identity that are the focus of much wider political and social argument nationally; this therefore makes it even more lamentable that it is not at the centre of any proper analysis of care delivery in the UK.

What in fact occurs – as alluded to above – is that those who are acutely aware of the dysfunctions that race causes in care either keep shamefully shtum or adopt a Pollyanna-ish approach that declares there is no issue surrounding race in care and that even if there were one it would all be ironed out in the unified pursuit of better care delivery.

Anyone who actually believes such evasive BS – and after all I have discussed about adult care being dead last in priorities in the modern care field – frankly need their heads testing.

Instead, and based on personal experience, I believe race and racial issues mar much of care delivery and contribute to the more toxic aspects of the care environment – it arguably contributes more to the division of care teams, the failure of structures of reporting and priority setting and the collapsing of morale than any other issue bar economic considerations. This makes a proper and clear eyed discussion of race in care not just overdue but a major priority. It is my aim here to redress some of that imbalance.

Let me start by making some bald statements that on the face it may appear paradoxical.

Care in itself – both in its administration and in its constitution - is inherently racist; in fact it is one of the most racist environments that I have ever worked in.

Care is a tinderbox of racism as it occurs on a person to person or CW to CW level – this is because it pits a plethora of different races, nationalities and cultures side by side with white British working class carers – a class (white UK national working class) shown in poll after poll to have the gravest issues with immigrants and those of other races.

Care involves some of the most racially diverse workforces in any sphere of employment.

White British born carers often are sensitised and responsive to diverse cultures way beyond what any broader poll of the prevailing attitude toward other cultures that their social or class origin predominantly would suggest – that being the white British working class.

Care exposes some of the less noticed and appreciated aspects of racism that exist in the UK today.

Without overseas workers it is my personal opinion, and backed up by a certain amount of empirical evidence, that wages would rise from the minimum in care.

And finally –

Racial issues are a significant barrier to the delivery even of safe care, never mind any type of quality care.

In the following I will expand on all of these statements to prove their veracity through my experiences but further than that I will look at all of the most controversial aspects of care as it is affected by race.

Some, no doubt, will be offended by my conclusions – the truth is I may be wrong in some or all of them, others may offer another point of view that may be equally valid – but that is the point of my raising these issues – to provoke a discussion that needs to take place because I wholly, totally and most vehemently disagree that these are issues better left not discussed – failure to discuss these issues is, as I noted at the end of the last chapter – the most heinous form of cowardice, yet it is more than that – it is a dereliction of the duty that any carer or medical professional should have - and that is the primacy of the safety, comfort, quality and responsiveness of care to those we support. If an issue of such swelling importance as race in care is considered beyond the boundaries of acceptable discussion then we have not really progressed at all over the last 30 or 40 years.

Finally in the following discussion I have tried to be scrupulously honest about my own conduct and views as well as regarding my experiences and observations, that may make it an uncomfortable read but at least it will be the truth and hope it will be received as such.

Think you aren't racist? Two weeks in care will find you out, after that you'll know one thing, everyone's a racist to some extent or another, including you. More to the point those who claim to - "not have a racist bone in my body" or, as justifications for actions and words – "I have worked with lots of different races and never had a problem before" (both of these are actual quotes) are some of the worst exponents of racism and of the worst forms of racism – the insidious and veiled type.

After two weeks in support work I thought I might be a racist, after two weeks in care for the elderly I knew that in order not to become an out and out one would take a huge conscious act of will.

That statement is intended to take you back a bit but it may help if I explain what I conceive of how someone who is a pure non-racist thinks. A non-racist looks at people not skin colour or nationality or creed, who never thinks about what background someone comes from when describing them and who views flaws and foibles first and foremost as human traits, not ones tied to culture or race.

I fail on all these counts but no matter, I don't know of anyone who succeeds in all of them either. The point is that all of those of us who would like to think of themselves as a non-racist are just plain wrong, if they labour under this misapprehension it means that either they have never actually been tested in the crucible of a multicultural working class environment or are born liars to themselves.

The above description of a non-racist is in truth an ideal – a worthy one and one worth aspiring to but nevertheless an ideal that can never be replicated; to become such an individual I would suggest would take a whole new paradigm of thinking so alien to our make-up that to even think of it is almost beyond the scope of our comprehension. This is because our whole perceptive mechanism is based on difference as a way of apprehending the world – at its simplest level it is – you are you and you are not me; at its racially most contentious it is – you have different skin colour, ethnicity or nationality from me. The more marked the difference – the most marked being you are you not me – the more immediately we reach for it – thus our whole perceptive mechanism is constructed around finding and identifying difference to understand the world and objects.

This prolix description is needed because we need to be clear what racism is, and, to me it is not simply discrimination – although I will use the term racism below to describe discriminatory or derisory language and actions – it is a mind-set. Thus in order for a supposed non-racist to be non-racist takes a positive act of thought – you perceive difference then have to actively mentally overcome it to see the person behind it and to divorce all their actions and words from absolute identification with difference other than they are another human being. Thus even when being scrupulously non-racist we are in fact practicing a form of racism in that we do not instinctively seize on the human but the difference.

The leap to reach a state where we simply see another human and not their overt difference is as impossible to grasp as non-existence – as we live we cannot conceive of not existing, therefore it is beyond our perceptive abilities, so too a world not based on difference.

The problems in what we classically call racism though are not based on such subtle distinctions, not viewed through the lens of different and equal but instead based on – at best - different and so very different – meaning there can be no common bond – and at worst - different and therefore inferior. Both these attitudes are classically racist, the latter more obviously so as it is this attitude which produces all the baleful effects of discrimination we are sadly familiar with – intense dislike, derision, disavowal and discrimination.

However the first is, in my experience, just as damaging as the second. This is because it seeks to hide itself behind a mask of respectability while emphasising difference; it creates divisions, an us-and-them culture, a process where working relationships barely function because "the other" - to borrow from philosophy – can never be properly known and therefore has to treated with caution and distance. Both these forms of racism are endemic in care, the above, more insidious form of racism, will be dealt with shortly, but it is the former – the more obvious and explicit - we will turn to first simply because it cannot be ignored.

Care is the most cosmopolitan environment I worked in, in my time in care I have worked with French, Germans, Spanish, Dutch, Czechs, Slovakians, Slovenians, Polish, Greeks, Italians, Moroccans, Somalis, Kenyan's, Nigerians, Chinese, Thai's and Vietnamese. At any one time the workforce in every care environment that I worked in was between a third and a half foreign born, some care home owners too were non UK born individuals and the majority of the nursing staff were non UK nationals. This aspect fascinated me as I learnt a great deal about other cultures and backgrounds from those I worked with. Care is like that, long shifts and the need to work intensely and closely with other people to achieve decent care outcomes produces very rapidly relationships that in other work fields or social environments would take many months, perhaps years, to produce. Within half a dozen shifts you will most likely know someone's background, their status, a potted history, their likes and dislikes and even their sexuality and belief systems. This is not through any loquaciousness or – as many people may think – long hours spent talking rather than getting any caring done (a most common refrain from relatives) but simply because even the most guarded of individuals will find it difficult to maintain a total reserve as tiredness or simply the endurance of time necessitated by 12 hour shift patterns, lowers all natural reserves of even the most socially diffident. I would class myself as anything but a chatterer yet the people that I worked with always ended up knowing more about me than I would normally have discussed with strangers, simply because I had to spend so much time cheek by jowl with them.

However this process of what I would call "social forcing" as well as pushing people together can also easily accentuate gaps and rifts. I too have seen a simple disagreement over something trivial escalate into outright hostility. Therefore introduce differences of race and culture and you already have potential fracture lines running right through the workforce. In the best cases – and the ones that observers of care like to "cherry-pick" as examples as how race *does not* affect care, but in fact enhances it – it produces the effect that I alluded to above in one of my statements – "White British born carers are sensitised and often responsive to diverse cultures way beyond what any broader poll of the prevailing attitude toward other cultures that their social or class origin predominantly would suggest" – the fact that CW's are forced into close social proximity with cultures, races and nationalities who they would not normally mix with – and toward whom polls would suggest that they would have – at best - an antipathy to – and at worst - an open hostility to - creates a cross-pollination that enriches the lives of all workers. Sadly though this occurs only in the – however substantial - minority of cases; what, in the normal run of things, happens is that the fault lines that already exist are further exposed and accentuated.

To put some definition on the scale of the racism problem in care let me give some brief examples of how I have heard CW's describe other CW's of another race or culture – "a baboon up a banana tree" "a monkey" "an ape" "a terrorist", "a sambo" "a Pikey/Gypo/ Refugee", "an 'African'", "a slant", "a slope" and "sooty" - none of these were uttered as an aside or even in close company – not that that would have made any of such phrases any more acceptable – but were uttered openly in the hearing of others and sometimes of senior managers. Often such "remarks" were thrown out in the course of idle conversation, but what was infinitely more shocking to me – and the nub of the matter as to how racial issues can cripple care – was that they were just as frequently uttered in the course of a professional conversation and in reference directly to other (absent at the time) care staff. For example the very first comment highlighted above – the "baboon up a banana tree" epithet was actually said by a senior carer in reference to another carer during a shift handover and in the presence of two senior nurses – one the deputy manager, both of whom laughed at the comment. I recall being half stunned when the senior carer actual said the words, and, for a moment, I was unsure if what I heard had actually been said or if I had misheard. When one of my colleagues actually called the senior carer out on this – challenging the acceptability of the comment - they repeated it with more force producing more gales of laughter from the senior nurses.

To me this routine racial abuse was unbelievable and shameful, and I am not easily taken aback.

This was not the first time I had heard racism in action of course – in the course of my many jobs I have worked with the most vehement of racists for whom the epithets "nigger", "coon" "black bastard" "Paki" or "Raghead" were not seen as a transgressive and deeply offensive terms but as part of habitual discourse on matters of race. However – and not to excuse the language or the underlying mind-set it betrayed - such terms were uttered in what I would describe as "wild west" jobs – semi-legal, self-employed, heavily manual and racially mono-cultural – that is white British working class - environments. In such instances there were neither potential professional sanctions hanging over them for using such offensive terms, or the potential of directing them at an actually present person.

In care though this was the first time that I had not only heard such pejoratives uttered in a corporate setting that had *on paper* a seriously worded and forceful anti-discrimination policy but also it was the first time that, in a professional environment, I had heard this abuse aimed at specific persons.

I was soon to become aware that this language was an extension of what many CW's actually believed that – that care was full of foreigners who were at worst totally incompetent and at best took up jobs that "English" people should have. Although on the surface there appeared some kind of unity this only served to make the undercurrents more shocking when exposed, for while actions aimed at attacking or persecuting non-UK national carers were never overt there was a more deadly rip-current that often made them disproportionately subject to difficulties in their work environment.

Most commonly these actions were the reporting of non-UK workers for "offences" veering between the petty and the serious; a couple of examples will illustrate what I mean.

One Nigerian care worker was repeatedly sanctioned on the say so of other CW's for a variety of small matters that were inflated into serious complaints. One involved a woman who had a propensity to cover herself in faeces whenever she had a bowel movement, the speed of which could – if this isn't appropriate – had to be seen to be believed.

I myself had supported this particular resident and discovered in the matter of 10 minutes they could easily smear faecal matter all over herself and her bed. One morning this CW had been left totally alone on an EMI unit with 10 very challenging residents just prior to a shift change. She duly clocked off when relieved by day shift only on the night after to be hauled into the manager's office to be given a verbal warning because this particular resident had been "discovered" by the day shift in a filthy condition. No mention was made that she had been alone and without help for the 15 to 20 minutes before she finished her shift nor had any consideration been given to the fact that it was well known that this service user *could* cover themselves in less than half that time.

What I believe was the racial subtext of this incident was crystallised to me a couple of days later when the same set of circumstances happened to me - I had clocked off believing all was well only to told that the next shift had discovered the same service user covered in faeces – the critical difference was I was not alone on that unit and so myself and my fellow CW had both been under the impression nothing was untoward, however no action was taken against me or my fellow CW despite, if anything, our infraction being worse because at least one of should have picked up on the matter. To me and my fellow CW it appeared crystal clear that the only difference was one of race – the black CW had been verbally warned for "neglect" whereas two white British CW's had had no action taken against them for exactly the same circumstances. To make matters more acute there were several times when I had taken over from the shift that had made the complaint against the Nigerian CW only to find this same service user in the same condition that had provoked the verbal warning for the Nigerian CW – the service user covered in faeces. When I reported it – numerous times – no action was taken – this shift being totally staffed by white British CW's.

This same CW came under a further opprobrium – and a written warning – after a resident fell on a night shift and had apparently been left on the floor for some time. The odd thing here was that she was working with another – white British - CW who – when the incident was exposed - blamed the Nigerian CW for not checking on this resident when asked. There was no way to corroborate this assertion and it came down to one person's word against another, however the White British CA threw in for good measure that the Nigerian CW was impossible to work with due to being inflexible, unresponsive to requests and intransigent. For the falls incident the Nigeria CW was sanctioned but the White British CW had no action taken against them *despite them both having an equal duty of care to keep this resident - who had a long history of falls and was therefore noted as being a falls risk - safe.*

I worked with both CW's and, as in the latter example, I found it was true that the Nigerian CW did have fixed ideas about how to work and regarding time management issues - no amount of appeal, persuasion or request could deflect her from what she had in her mind to do when she wanted to do it. However the exact same was true of the White British CW who was an old school carer who had been in the care field – as they never tired of telling everyone – for 20 years and who was absolutely certain that her way of doing things was not only the *right* way but the *only* way to do them.

To even the most cursory of investigations it would have been as apparent to anyone, as it was to me, that here was a classic clash of two very strong personalities neither of whom was willing to accept a different way of working, yet this very obvious fact was ignored (I cannot believe it failed to be noticed unless the management was more dim than the little credit I gave them) in the headlong – and racially motivated - rush to judgement.

It was a similar story with another CW I worked with, this time an Asian one. This CW was one of best carers I worked with in regard to how he interacted with service users and how sensitive and considerate he was to their needs. Their one flaw was that they were dreadfully absentminded and therefore needed benign oversight of the sort that offered encouragement and guidance. He worked night shift and travelled some way to do it (a point we shall return to later in regard to overseas workers) – on a night it took him 2 hours to get into work, and the following morning it took a good 90 minutes to get home.

One night working a floor alone - where one night worker was assigned to 15 service users - a service user had a fall. The nurse on duty attended and there were no serious injuries. Despite never having being shown this CW had to fill in an accident report which he did, unguided, and submitted it. On the morning he went home believing they had fulfilled all the administrative obligations only to be called by the home to say he hadn't filled in a *company* incident form. I personally – who had worked there much longer than this CW for the same company, and had filled in accident reports before - had never known or been asked to fill in a *company incident form*. Quite understandably, as he was half way home, the CW asked if they could fill it in on that night only to be told forcefully in the negative and that he must return to the home to fill it in. This he duly did – filled in the form which took all of 15 minutes - then set off on his way home *again*. The incident form turned up a few days later on a different unit and lying among assorted paperwork – *it had not been filed or put with the accident report,* in fact it had just been left lying around – totally superfluous to requirements.

The question then is why was this CW dragged all the way back into work to fill it in? Overzealousness? It seemed difficult to credit as never again was I or anyone else I knew *ever* asked to fill in such an incident report in such circumstances.

Lest it be thought this evidence of discrimination is rather thin, this CW also incurred a written warning a short time later. He was working a floor – again alone – where one service user, due to having had half their bowel removed - was prone to heavy faecal soiling that was more or less continuous. Trying to get this service user clean was extremely difficult as even when cleaning them they would at the same time still be soiling themselves. The CW in question attended to this service user in the morning after getting them up and there seemed no problem. When they returned on the night though they were given the hook into the managers' office and made to agree to a written warning because they had left the service user and their room "covered in faeces".

I did not know this CW had been given a written warning until much later – 2 weeks or so – otherwise I would have intervened as I had been in this service users' room *after* the CW in question had attended to this service users personal care. The service user appeared clean to me and the room too – certainly it was not "covered in faeces". While it was possible some marks had been left – the volume of faecal matter and its consistency made transference from gloves to other surfaces almost unavoidable even with the best of attention to hygiene measures - and in past in supporting this service user I had left marks on different surfaces but luckily I had spotted these and cleaned them up – it most absolutely did not match the apocalyptic references made to the condition of the service user and their room.

Further, and in a reprise of what I have already said above - I had come on shift in the past – taking over from the shift who had made this complaint against this CW - to find the same service user in the most appalling of conditions – covered from waist to ankle is faeces and having been left in that condition for some time as much of the faecal matter had become dried on. What is more I had in the past been into this service users room to find used wipes and continence pads on the floor of the bathroom – again on making a complaint about this nothing was ever done – certainly no written warnings were handed out – this despite these infractions being much more serious than that of this Asian CW. The only possible explanation for the outsized sanction against them was their race – how else to explain the lopsided treatment of different CW's.

These are just a few examples – I could go on, believe me – for instance a Ghanaian CW who was extremely good at caring who was constantly made the object of complaints and bitching among care staff, a Polish worker labelled as lazy for no apparent reason, a Thai made to attend a training session under pain of disciplinary even though her 2 year old child had chickenpox – while others were allowed to doge it on the flimsiest of grounds – the list is endless.

I have no idea of the mental or emotional cost to these workers but it must have been high – is it any wonder that they may become demoralised, de-motivated and disinclined to put in the extra mile that care always involves?

Here we are presented with what I believe is the immediate and visible consequence of discriminatory practice rather than the initialising factor in it – having been so run down by management and fellow workers is it really a surprise if non-UK CA's became less willing to work hard? After all, if the reward for working hard is unjust actions, whispering campaigns and disciplinaries, what is the point in pushing on at work?

This produces a feedback loop of the most destructive sort – rundown and ground down by so called colleagues, discriminated against CW's are inclined to do less and less, and the less they do the more they are labelled lazy, the more they are labelled lazy, the less inclined they are to do more than the bare minimum, and so on – a self-fulfilling prophesy and reaches its ultimate absurdity, apotheosis and irony in the fact that all this reinforces the prejudices of the racists – they believe they have been vindicated in their pre-emptive judgement of foreign CW's as being predisposed to laziness.

To restate, and to emphasise: **the fact of racism in care is, I believe, is the main and major cause of a widely held perception in care – that non-UK workers are shiftless and incompetent**.

Black African workers most particularly bear the brunt of this racism, with the result that the most widely held, voiced and actioned stereotype in care – almost always without good foundation – is that "they" or "these people" (how far down the racism pit are we descending here?) are inherently lazy, incompetent and useless - all this solely due, in most cases, to the process outlined of a self-reinforcing, a self-fulfilling prophesy.

However CW's being abused and victimised by management and co-workers is only one strand of racism, and, bad as it is, it is not the most pernicious kind, the worst happens when closet racism runs up against non-UK nurses.

When authority runs up against race conflict is almost certain to arise and these conflicts are often thrown into much sharper relief than the more covert racism operative at CW on CW level. There is a perception in care that overseas nurses are less competent and therefore less worthy of notice than UK nationals. I have observed this time and again – instructions not followed, requests unanswered, support declined. A major part of this comes down to language though, and while this doesn't excuse the underlying racism it offers a snapshot as to why it occurs with such vehemence and frequency.

The barriers to communication are often marred by the fact that for all overseas nurses English is their second language, often times this is heavily accented and littered with mangled colloquialisms and malapropisms (more on this later) – this gives the *impression* of a lack of competence which actual practice gives the lie to. More than this though is that in the provinces there is also the barrier of accent and idioms that often serve to confuse nurses struggling to understand what they are being told. Throw in the fact that medical terminology fogs the issue further and you have a recipe for almost constant misunderstanding, misinterpretation and mistakes.

I worked with one Thai nurse who had also worked on the continent so mixed with the lilt of her native tongue was also a broad mid-European twang all rolled up with often amusing mangling of phrases. To many first encountering her the grave error was made that she really did not know what she was talking about – wrong, very, very wrong; she was in fact one of the most acute, professional, knowledgeable and unflappable nurses I have ever worked with – no matter what the nationality - this was combined with a steel edged certainty of what she expected of CW's working under her direction.

Hers though was not a way of confrontation, but of a collegiate spirit where her professionalism was expected to be matched by that of CW's The result was that those who did not know her well and who thought her language betrayed a poor grasp of care matters often found themselves swiftly disabused – she loathed laziness and would not be backward in coming forward in telling shirkers what she thought of their attitude.

As a result many (white British) CW's did not like her and numerous (spurious) complaints were made about her. While, as noted above, these complaints, however unfair, however unfounded, would be enough to do for a CW, nurses were more prized – employing and retaining good nursing staff is extremely difficult in the care sector so when one is employed they gain a degree of traction.

Not being openly discriminated against though is not the same as being supported. This nurse, in answering silly complaints made against her and in making valid complaints of her own was *never* and I mean *never* taken notice of. Finding residents soaking wet in their own urine or covered in faeces or shut in their rooms without stimulation were all complaints she had to make numerous times yet nothing ever came of them. Worse was when defending herself against the false claims she was not supported by a management that, rather than stepping on the issue of discrimination and resentment, chose instead simply to go AWOL on the manifestly visible problem. Over time this failure of support and the constant complaints wore this nurse down and she finally resigned to go to work in the NHS, fed up with being ignored and disempowered.

Arguably this nurse also fell afoul of the other problem that language often creates and this is a bluntness that is unintentional.

While speaking a different tongue can be mastered broadly, the nuances are infinitely more difficult to grasp, as anyone who has falteringly tried to apply a scanty fluency in another language, only to see the person they are talking to put on the back foot because we have asked something devoid of customary politeness, will be all to aware of.

(As an interesting aside of the nuances and mental disconnects that occur in using a second language see - http://www.economist.com/news/science-and-technology/21602192-when-moral-dilemmas-are-posed-foreign-language-people-become-more-coolly)

In a position of authority it is sometimes the *way* CW's are asked to do certain tasks that sets them on the back-foot more than *what* they are asked to do. This is enough to cripple a working relationship right at the start – anyone would respond poorly to what they see as very direct "orders" instead of requests from someone in a supervisory position.

Take the following statements.

"I have just given Mr X his medication and I think he needs changing."

Or put another way –

"I'm doing my medication – you must change Mr X"

The same action is asked of a CW but the second is much more brutally direct, yet to a speaker of English as a second or third language both mean much the same and would wonder what provoked a poor attitude and pushback if they employed the second phrase rather than the first – thus are misunderstandings and poor relationships born. This would be an issue in any profession but in care, taking into account the especially thin skins that most CW's have – and much of the emotional baggage they carry with them into care – this can be terminal to a working relationship. But more on language later.

Having said all that though make no mistake - the ground on which such misunderstandings are founded is one of racism, a simple understanding of the lack of nuance inherent in second language speakers would be enough to defuse offence yet none of this is forthcoming – White British CA's *expect* a difficult relationship so they are bound to find one. The same goes for competence – if someone instinctively and chauvinistically expects you to be incompetent they will sooner or later find reasons to prove this is true – as the above example should show racism, although rarely bubbling to the surface, is endemic in care.

I have highlighted above the more clear-cut and what I would call "traditional" manifestations of racism – that of white British workers against those of other races, cultures and nationalities, but before moving on to look through the looking glass the other way (yes there is another way) something must be said of the casual racism that exists every day in care.

Having noted what I have said about difference at the beginning of this chapter it must be added here that while difference does play a factor, how we articulate difference – whether it achieves a primacy over the individual or whether we consciously or unconsciously place the individual before any difference, is to me the defining characteristic that is displayed in casual shows of racism.

An example should highlight what I mean – working on a night shift during the winter I came into handover and the room was stifling, although the windows were open they were all (as all windows should be in care homes) locked to only open a couple of inches, the heating was belting out at what must have been full blast and even though I was chilled by the walk into work, within minutes I was sweating, as were my work colleagues gathered. The Deputy manager came into the room closely followed by the (Nigerian) night nurse (as it happens an individual who, as will become apparent shortly, I, and others, singularly failed to get along with; and whom therefore I could be said to have had little sympathy for – which adds piquancy to the following observation). The night nurse commented on the heat in the room to which the deputy manager responded – and I quote verbatim –

"How can you be hot you're an African."

It's hard where to start with this asinine and offensive comment. The fact that the night nurse was viewed as an "African" first and not a person who found the environment hot, the boneheaded ignorance implicit in the phrase - that "African's" don't get hot -, the complete needlessness of bringing race into the equation or finally giving the appalling example to a gathered workforce that bringing matters of race into any issue was permissible.

I personally was stunned that anyone, never mind a deputy manager, would gratuitously make such a casually racist comment *and* make it in front of a workforce who were subordinate to the nurse. I was not the only one taken aback, so too were my fellow (white) colleagues, however the nurses' reaction was more pointed – I quote the following exchange, again verbatim, to give the flavour of the minor altercation that followed and the hole the deputy manager kept digging for himself.

Immediately following the comment the nurse responded –

"That's a racist comment."

"No its not."

"Yes it is racist."

"I was just saying you should be used to the heat."

"You are a racist."

"Me? I haven't got a racist bone in my body."

At this the night nurse sat down, obviously simmering. Although voices were raised the to and fro never became overly heated but was more affecting due to that fact – offence *had* been caused - and needlessly so.

The plain words are striking enough but they do not convey the tone – the initial comment was made in a mocking fashion, the subsequent comments of the Nigerian nurse betrayed not anger but a degree of hurt, and the deputy managers' lumbering attempts at self-exculpation were also delivered in a patronising and belittling key, the whole thing giving the impression of a senior staff member getting a rise out of a more junior one and in front of an audience (we CW's) to make it all the more effective.

Some will judge this whole exchange as a storm in a teacup or indeed as evidence of the oversensitivity to race that "disfigures" common discourse, yet it *was* telling of *something* – the reduction of someone to origin, race and, ultimately, skin colour. Would someone make a comment based around a white nurse being able to stand cold better, because they are white, or European, or British? Maybe in Africa you may respond. Fine but we were not in Africa, we were in the UK and in the UK there are certain standards of professionalism and even-handedness that we profess to hold and upkeep.

The comment, like it or not, seriously undermined this nurses authority over the workforce – the casual manner in which she was reduced to a skin colour and background could only diminish her standing – if a deputy manager could treat her in such a fashion what message would that send out to workers? Even more than this though was the attitude it betrayed – one of dismissiveness and disrespect.

This was just one event in several I could add to here and is indicative of the way the race issues are handled, at least in private sector care.

However such racism, although the most prevalent, was not the most virulent that I encountered.

Perhaps I was totally naïve to the point of stupidity but I always instinctively associated racism as a white on coloured thing, it never occurred to me that great tensions could exist between other different races.

By far the most brutal I encountered was that perpetrated by black staff members upon Asian ones – whether of the Indian subcontinent, Middle East or Far East. I saw CW's from Asia treated with utter contempt, subjected to constant small criticisms of their work, made to miss breaks, reported for minor infractions (what an irony that one) and, ultimately, forced to leave.

One CA I worked with from the Philippines was a conscientious and solicitous worker, got on very well with all the service users, was efficient, able and affable, yet they were almost always in trouble with the nurses of colour; in trouble for not propping a service user up properly in bed, in trouble for taking too long in conducting personal care, in trouble for entering another service user's bedroom alone (this was to do mandated hourly "breathing" checks and to ensure the service user was dry and clean) in trouble for not filling in paperwork in a satisfactory manner – although when I checked – at the harassed CW's request - I could see nothing wrong – in trouble for taking their breaks at "the wrong time" and in trouble for not wearing the proper uniform (this was during the summer when he removed his tunic because of the heat and wore a t-shirt instead that was white and smart); basically he was in trouble for even breathing.

This constant pettifogging of their work eventually wore this CW down (I was soon to know what this was like, more of which below), he told me in desperation that he could do nothing right and was always being carpeted for things that were to a large extent minor or at best wholly subjective – quite *how* you can prop someone up improperly is beyond me. They also asked me to observe their interactions with coloured nurses and I was shocked by the manner they were referred to, the attitude shown to them and the torrent of complaints and moans. I asked the CW if he was going to complain but they just shrugged their shoulders – what was the point – they asked rhetorically – how could he justify himself against the complaints of nurses. Eventually the nurses in question went to the manager of the facility and stated they were not prepared to work with this CW, and, as there was no alternate post available the CW was moved on to another home in the company. As they were the only Asian working on shift at that time (substantially more have since been recruited bringing its own problems – see below) it is difficult to see these nurses' actions as anything other than a concerted effort to have this CW either dismissed or moved on - first trying to wear the CW down with the constant complaints and, finding that that didn't work, eventually forcing the issue by refusing to have him in their team.

I found this episode shameful, shameful in the treatment dished out to this largely inoffensive worker and doubly shameful in the fact that it proved wholly successful to the nurses' aims.

However this episode was typical of how Asian workers were treated, as I have seen other Asian workers sent to work in the kitchen or laundry where "they would be of more use" – yet another lazy racial stereotype that "they" were better suited to cooking and cleaning than care.

Once more this form of racism goes on "under the radar" largely not noticed, underreported, and, even when it is reported, those complaints and grievances ignored. This produces a vast swell of resentment, especially as ever growing numbers of Far Eastern CW's are today employed in care with many private employers actively recruiting abroad. These Far Eastern CW's more and more find themselves under the authority of coloured nurses and other CW's who go on to perpetrate this racial abuse in an endemic fashion.

In one home that a colleague I knew worked in these tensions for once broke the surface, but only after the number of Asian CW's had reached critical mass; one coloured nurse had so alienated a workforce that they effectively mutinied, all of them simply refusing to work with this nurse. The management could not afford such a breakdown so they removed the nurse from frontline nursing duties then shifted them to another part of the service. This though was an isolated incident and an exception.

The joke was this nurse was the same nurse that "engineered" the removal of the Philippino at the home where I worked – singular evidence, if any more were needed, that such racism is not so much missed as tolerated.

Two more important points need to be covered in regards to race. The second is a personal story which I leave the reader to decide the nature of, the first though asks difficult questions about the employment of overseas workers and especially nurses in care for the elderly.

Two things need to be clarified here and one reviewed that, although they may be uncomfortable to hear or read, are I believe important to make simply because they *are* uncomfortable. Firstly let me review a point I raised at the beginning of this chapter – that the almost endless supply of cheap overseas labour in care has the effect of holding down wages to the minimum throughout the private care sector.

Not everyone is suited to care; you have to deal with all manner of bodily fluids, be subject to potential and actual attack, work long and unsociable hours and are expected to meet various and oftentimes conflicting priorities and expectations. Against such a grim parade of facts you have to question *why anyone would work in care*. The personality types have been covered in a previous chapter but in the main it comes down to people work in care – at least the good carers – *because they want to*, because, through all the most difficult aspects of it, care can produce and display the most wonderful aspects of existence, can produce moments that make you feel happy and fulfilled that would be barely possible in any other type of job and brings you closer to people at the most basic and human level that we all share.

Such people are - by the very fact that they show an unusual degree of skill-sets - from being cleaners of bodies to a willing ear, from defusing confrontation to slowly and painstakingly building relationships across almost insurmountable barriers - a finite commodity – there are only so many people willing to put in the hard yards just to feel the wonder. According to supply and demand a premium should operate in the care jobs market – the fact that it doesn't is wholly due to overseas workers.

This though is not a nativist appeal or racist attitude of "send the buggers back" or "British jobs for British people," it is just a blunt recognition of the facts – that at least 20% of care workers in the UK are non-UK residents (this number comes from figures from King's College London but is almost certainly an underestimate, the real number being closer to 30% and in some regions markedly higher)

Almost all of the overseas CW's I worked with were excellent carers and dedicated workers, more, much more, than the average British worker. Heroic levels of sickness prevail among white UK-national CW's yet overseas workers, in my experience, rarely missed a shift or failed to work hard throughout it.

One example will tell you all you need to know. One shift I worked a white UK-national worker (who was either the most sickly person this side of death or the most malingering) phoned in sick only 90 minutes before the start of the shift, unsurprisingly no one wanted to come to cover the shift so an agency was called. With, by then, only an hours' notice, an overseas worker travelled from *25 miles away* to provide cover; 25 miles to cover a shift for which a good part of their wages would be eaten up by travel costs – why? Because they were desperate to work.

Enough said apart from if UK-national workers showed even half the work ethic that overseas ones display care would function far and above its dysfunctional incapacities of today.

Against that background is it any wonder that overseas workers may actually be preferred to UK-nationals? Yet the point remains – as long as there are cheap and reliable workers on hand it will be impossible to drive up standards and pay in care.

This may seem like a contradiction to what I have just said about the reliability and work rate of overseas workers but in fact it is not.

If the aim of the care sector, private or state funded, is to have care delivered by a professional – in all senses of the word - workforce that provides a career based on decent remuneration, constant up-skilling and one where carers were enrolled in and represented by a professional body that held them to the highest standards while arguing their corner, then it cannot be based solely on cheap labour – a race to the bottom in wage terms is antithical to one driven by quality considerations.

This point is so obvious is seems ridiculous to make. For example if you ran a business with a large turnover and high volumes of trade would you employ an accountant simply because they were cheap? Would you employ a Chief Financial Officer who was not held to professional standards? Would industry support a key sector of financial oversight even though its members were effectively free from any basic professional qualification? Could anyone in retrospect, and taking these factors into account, complain if the wheels came off due to mismanagement of capital or fraud? Yet this is the environment that care dwells in. These points have already been made elsewhere but they have a special piquancy when applied to the large and swelling band of overseas care workers.

I am not suggesting that overseas workers lack professionalism – as the above shows they in fact show more professionalism that UK-national CW's - no, the problem is that contrary to popular opinion, most overseas workers have no intention of staying in the UK permanently, instead they are here to work as hard as possible for a period of time in order to send money back home to either provide capital for their own future ventures on retuning or to support an extended family, or even just to provide a nest egg for them fall back on; as a result they are not in main interested in accumulating qualifications or new skills because they will not, in their eyes, need them for the long term.

All this means care will not only be kept artificially cheap on the labour side it will also never lose the tag of being the preserve of "low skilled" workers.

The more difficult and awkward facts comes when we turn to clinical staff. In contrast to CW's overseas nurses, according to my experiences, are of vastly varying quality. Some I have worked with were the best of *any* nursing staff I worked with, period; others were poor to the point of being a liability. This variability is pretty worrying for most observers in care, seeing as private care provision, especially care for the elderly, unlike in acute clinical settings, has little or no peer review or senior nurse oversight, therefore mistakes and errors go uncorrected and accountability is lacking, in fact it often takes someone to die due to negligence or incompetence before failing nurses are spotted and made liable.

These issues are not restricted to overseas nurses – some of the biggest mistakes dropped that I have known – including a death – have been by white UK-national nurses – however the prevalence of poor nurses is greater among overseas nurses. Among the failings I have personally witnessed – the failure to diagnose an broken wrist in a service user despite them complaining of extreme pain, failure to diagnose a broken shoulder, failure to diagnose a broken hip, the splitting of a service users lips by a nurse forcing tablets into their mouth with a metal spoon, failure to act on signs of internal bleeding, failure to diagnose a heart attack, leaving open Grade 2 bedsores on the sacrum area undressed and so becoming infected by faeces, and routinely missing medications and errors in drug recording, including controlled drugs. And these are just the highlights; I could go on, and on and on. The question here though is with such errors how can a care provider support such nurses? Surely they present a grave liability corporately? In fact they do but only if such errors are properly noted as such. In fact in every case noted above the mistakes have been covered up, paperwork has been altered, accident forms noting the dates of accidents have gone missing, and MAR (Medication Administration Record) charts "manufactured" after the fact.

This still does not answer why though - why retain poorly performing or incompetent nurses?

The answer, in the main, is that they do a lot of overtime and are willing to work antisocial hours; in fact they are willing to work *any and all hours* (this itself is a problem we shall return to shortly) and, of course, they are cheap compared with what a UK-national nurse would expect in terms of pay and conditions.

Care providers therefore have a classic conflict of interest, if they have one or more nurses willing to work any shift and at the shortest of notice they will need to support fewer staff – having nurses willing to work 7 days a week 12 hours a day means that they need employ only one nurse instead of two or three. For the same reasons were they to fire poor nurses who do a lot of shifts they may have to employ two or more nurses to replace them with all the on-costs associated; therefore it makes financial sense to cover up mistakes and errors.

The problem of incompetence or negligence though is fed and compounded by the hours that many overseas nurses work, in fact nurses who are not in fact incompetent and negligent can be made so by the hours they work. As I mentioned above it is not uncommon for overseas nurses to work 7 or more days straight of 12 hour shifts. It is simply inconceivable that such hours will not affect performance; where a nurse is incompetent or negligent in the first place what you have is a ticking time bomb waiting to explode in a serious and life threatening incident.

The hours worked by many overseas nurses does themselves, in most cases, no favours, as many – and in a curious add on to the racist stereotypes and attitudes noted above in care - mostly black African nurses – acquire an immediate tag of being lazy. In truth I have never met a lazy nurse of whatever colour (except white – 2 of the laziest nurses I worked with were both white as it happens) but I have met plenty of overworked ones - the urge to work ever more hours reinforces – but does not create – the notion of laziness associated with skin colour and race (that notion being fundamentally racist without needing any supporting foundation). The sad part is I have known nurses who were very good at their job but rendered totally incompetent by the severe hours they worked.

Too many hours worked also affects CW's and again is mostly found in overseas CW's - this too introduces problems of stereotypes and immediate reputation, the main difference though is that while CW's are vital to the functioning of care, their duties do not involve issues that may be dramatically detrimental to a service user – unlike nurses - who's broader area of clinical care leaves more scope for more - and more serious - errors.

Here I would like to point out again that incompetence or negligence is not tied to a race, colour or nationality, the points I have raised above are in relation to most overseas nurses whether they are East European, Asian, Mid-Eastern or African – skin colour and race have nothing to do with it – the variability is universal.

The final point I would move to is one tied to background and overseas nurses. Many of the nurses I worked with spoke better English that UK-national workers, however a good minority spoke with very heavy accents and mangled syntax. Of these there were a portion of whom I found it very difficult to understand myself, and the same went for my fellow CW's who often were non-plussed as to what they had just been asked what to do by such nurses. This was a problem for us but it raised a more critical issue – the fact of how confused, often hard of hearing elderly people with poor word forming or understanding skills were supposed to be communicated with when even those of us with all our faculties, good hearing and decent language recognition skills had real problems with understanding information passed on to us by nurses.

This issue of language has received, at last, the coverage it deserves more recently when the BBC highlighted the problems caused by the poor language skills of CW's – but notably stayed away from the same issue in nurses – a shameful omission.

http://www.bbc.co.uk/news/health-27295554

To give a flavour of the problems caused by poor or hard to understand accents and language I once witnessed the following conversation between a nurse and a 100 year old service user who was confused and disorientated by a pre-existing Urinary Tract Infection (UTI) and was complaining of pain in their torso –

"Do you have any constipation or is it angina attack?"

The service users looked back in askance at the nurse, plainly not understanding a word said.

"Is it pain of chest or abdomen?" the nurse enquired further.

Again a blank look.

"Have you had bowel movement in last few days?"

Again a glassy eyed response.

The nurse then left the room and advised me to keep a close eye on the resident and then left the floor I was working.

This conversation is verbatim, or as close to it as I could approximate in the immediate aftermath as I made notes of it soon after its termination, worried as I was that I may have an incipient angina attack case on my hands and wanted a record of what had preceded it. Luckily the woman was neither constipated nor on the verge of an angina attack but was actually suffering from the residual effects of her UTI.

This conversation is recounted here not to ridicule an accent or form of words but simply related to pose the question of how was this service user

1) Supposed to understand the technical terms the nurse was using

And

2) Supposed to decipher the words and their meaning through a very thick accent.

This is a very real issue, if a service user cannot communicate effectively how are they supposed to describe and articulate pain and discomfort. Imagine going to your doctor and not being sure you can make them understand what your symptoms are and then in turn not have a clue what type of remedy they are suggesting; in pain and frustrated is it any wonder that pain management and a lack of clinical engagement create more challenging issues than any other cause.

I am not for a moment proposing that anyone with a foreign accent be disbarred from care on those grounds alone, what I am asking though is - do we really consider the service users needs and requirements when recruiting staff? It would seem plain to me – and now to the Department of Health also - that a minimum of language skills be needed, for not just CW's but also nurses, working in care so that at least service users have a chance of making themselves understood and understand in return.

All in all and overall these cases - of whatever colour or race discriminating against whatever colour or race - I too felt shame welling up inside of me. Before I worked in care I liked to think that if I saw racism I would not stand idly by and do nothing; yet nothing - in all these cases I witnessed - is exactly what I did. Never once did I personally put in a complaint about the abuse I witnessed, never once did I offer firm support, never once did I step out of line on a point of principal. I could have said something, I could have registered a formal complaint, I could have offered to support someone else's complaint; I never did. I especially felt shame about the Pilipino CW as in looking back at his asking me to review his work, to witness the bullying and intimidation; I could see that implicitly he was asking for my support. I could have – should have – offered it to lend weight to any complaint but I didn't.

Why?

For exactly the same reason I did not step out on all those other occasions – I was afraid of putting my head above the parapet, of exposing myself as some sort of troublemaker, of being alienated from other CW's myself; and, in the last case, I was aware I was going to have to keep working for the nurses in question – and one in particular – the ringleader of the bullies - a working relationship that had started badly, then improved and then, as I shall show, went on to fall apart - if

I had complained it could have put *my* position in jeopardy. I chose the cowardly route and said nothing. This played a significant part in my finally becoming disillusioned in care, I was not the person I thought I was – not even close – and if I was not that person who was I?

This would be a question that would, over the next 2 years, be answered.

The full irony of my failing to offer help and support to the Philippino CA only came crashing in like heaving roller out of a ragged sea in the months that followed his departure.

I had worked with the one non-UK national nurse - whom had been the exponent of the victimisation of the Philippino CA - for some months. Initially I had got on well with them but soon experienced a rocky patch. I found this nurse very overbearing and very domineering however I must also admit that I reacted badly to their attitude. Therefore in many of the issues that arose as matters of contention and even argument between us I – rightly - shouldered the blame for. I perhaps was not as receptive as I might have been, not as alert to the requirements made of me and occasionally dismissive of guidance. In this, and as a subordinate, I accepted full accountability. Be that as it may it has to be said that I was not wholly responsible either – there is a subtle difference – in our many disputes I was responsible for my own actions but I was not accountable for situations not of my making. For instance it was very difficult – toward the impossible - to follow every instruction issued to me from this nurse as they sometimes were in contradiction to each other, sometimes they were totally unreasonable and at others just plain asinine. It felt like I was being micro-managed, that I was being told not just what to do, but how to do it, when to do it, and often in a fashion that I perceived was not always necessarily in the best interests of those I supported.

A very small example of this wider problem is instructive. I was working on a night shift on an EMI unit together with another CW, I was busy supporting a service user in having supper when the nurse came on the floor, I was in a resident's bedroom at the time and heard the nurse calling out the name of my fellow CW, I carried on supporting the service user and when finished came out of the room to find the nurse standing confrontationally outside.

"Did you not hear me calling?" she demanded of me. I replied that I had heard her calling for the other CW.

"Why didn't you respond?"

"You weren't calling my name." I answered.

"You should have answered instead of ignoring me."

I was going to point out that in future if she needed me I would suggest calling my name, or, if now any name that was called out was my responsibility to answer, what number and kind of names she would like me to answer to. Needless to say I didn't – I bit my tongue again (which by now was in metaphoric ribbons) and walked away before I said something I might regret.

I did wonder though how I was supposed to be a responsive and attentive CW when I couldn't be sure whether I was actually being "summoned" or not.

Things came to a head during a nasty altercation with one very challenging service user which I felt I was unsupported in trying to fix a situation not of my own making. Eventually I was given the hook by the deputy manager and spoken to regarding my relationship with this nurse. I was told to shape up or risk further action.

Dutifully, and sincerely - for, as I noted above, I examined my past conduct and found it wanting in certain areas - I made and apology in private to the nurse, promised to redouble my efforts and to move forward.

I did try to change but the change in the relationship really had nothing to do with me, although I may have been more receptive to minute instruction, I continued working in the best way I could with no marked change. The only change I could observe was in this nurses' attitude to me, although I did not do anything substantially different the relationship improved because – as far as I could see – her attitude changed to me.

This hiatus – lasting a few months - was only temporary, over time it started to sink back. Once more my every action was questioned, my every movement challenged, my every small departure from what *this nurse wanted* - which often was not necessarily the same as what a service user verbally requested, and, as noted above, was not always clear and sometimes plain contradictory - was greeted with the most astounding hostility.

After every shift I did what I always did - walked home and then went for a run – which was my routine post shift way of relaxing – and I used this time to look back and see if I had done all I could during the past shift, been of the best service to the service users I was supporting and if I had reacted to instruction in a constructive way. Shift after shift, week after week and, eventually, month after month, one of my main occupations during this timeout was spent in racking my brains to work out exactly what I was doing that was rubbing this nurse up the wrong way and too try to see if there was a way to fix it. This process took up more of my time and consumed more of my mental and emotional reserves than possibly anything else. I tried everything – and I mean everything – within my power to try to make this relationship work. Still the constant carping continued, or, in fact, became worse. Now, if I was supporting any of the service users with another CW and the nurse found something not to their liking it was *my fault* alone, and when I was working with a service user by myself then whatever I did was wrong. Two examples will hopefully show just what I mean.

One resident who was diabetic and insulin dependent often experienced dangerously low blood glucose (confusingly called BM) levels first thing on a morning so it was suggested by the clinical team that this resident was woken early and given something to eat. I tried everything, from toast to cereal, to keep their readings up – nothing worked and meant post reading I had to give them more cereal or more toast even though the resident was full and did not want to eat anything.

Eventually I hit on yoghurt; first I tried one, which the resident devoured with gusto, this too failed to move his BM's so, seeing as he was receptive to the yoghurt, I tried 2 – this did the trick, it was enough to raise his BM's without them skyrocketing – further testing throughout the day proved that this wasn't a too large a sugar hit as the levels remained up rather than slumping as would be expected if too much sugar had been introduced. More than this I had a further rationale – if the BM's did not move sufficiently I could still offer cereal or toast and so give a choice and variety to the service user beyond more of the same. As long as this service user had 2 yoghurts their BW's were fine. For two weeks this proved to be the solution, then suddenly the nurse said that yogurts were too sugary and I should give the service user cereal instead. I politely, and as tactfully as I could, pointed out not just that the yoghurts did the trick as far as the BM's went but also offered by further reasoning – that a choice could be offered if the BM's still proved too low. All to no avail. So I reverted to cereal – the result – you guessed it – low early morning BM's – for which I was harangued for; and so back to the beginning of stuffing this poor resident with toast and more cereal to get the BM's up.

It got to a point that the following shift complained that his BM's were falling because this service user did not eat their breakfast – not surprising seeing as that once I had finished plying him with cereal and toast that they were coming to resemble a cornflake or a dishevelled loaf of sliced bread. So, without informing the nurse, I reverted to the 2 yogurts – surprise surprise the BM stabilised. Of course the nurse asked me what I was giving this service user so I told her I was giving cereal and toast, to which their response was, and I again quote verbatim –

"I told you the yoghurts were no good but you wouldn't listen, you never listen."

The urge to come back on this was almost uncontrollable but, as I was learning to do – I bit my tongue and engaged my brain before my mouth – I just concurred with the nurse and agreed she had been right all along. The smugness of her expression was almost intolerable but I told myself it was worth putting up with as the service user ultimately benefitted from my dissimulation and it made my life easier.

This episode begged the question – did the nurse actually want this service users' BM to hit the floor so she could allege I wasn't supporting this resident properly with their diet? Did they want to create a situation that demanded her intervention to "fix" it? Was this another way to exert control – that she could effective "control" this service users' BM? Or - and this in my most troubled moments of post shift reflection – did she want to humiliate me, to see me admit I was wrong, to be wholly professionally prostrate before them?

Even though I dismissed this last thought as paranoid bunkum it kept coming back to me – the relationship between myself and nurse had been at its most placid, least fraught, when I had deeply apologised for things I may or may not have done – was this the way, the only way, that they could be satisfied?

Another incident just about finished my relationship as a work of collaboration and unity with this nurse. It was, as usual in matters like this, a very small thing but it just seemed indicative of the whole collapsing nature of the process.

A resident – in early stages of dementia but with capacity in decisions affecting their wellbeing on a day to day basis (excluding medication) - with a very poor sleeping record that we had been asked to monitor – recording how many hours they slept each evening - liked to be propped up on three pillows but, when they slept, they slid down the bed but otherwise were quite settled. Often this meant them adopting strange sleeping positions – their head tilted right back and snoring open mouthed. As always, and before they attempted to settle down for the night, I moved them up the bed and propped them up on their pillows. One evening I noticed, before the service user had fallen asleep, but when they were just "dropping off" - their eyes heavy, their responses slow and languid but otherwise aware and retaining their customary bluntness in speech - that they had moved back down the bed, I went into the room and asked the resident if they were comfortable – this conversation I noted afterward so is quoted word for word.

"Hello [X] you look in a pretty uncomfortable position there, d'you want me to move you up the bed?"
"Why?"
"So you are in a more comfortable position."
"I'm in a comfortable position now."
"Are you sure because I can move you if you want?"
"Well I don't want."
"So you're sure you're comfortable there."
"Apart from you disturbing me, yes."
Chuckling together as we always did at such moments I left the room. Half an hour later the nurse came up to "check" on me - I was filling in paperwork at the nurses' station. The service user who I had not long offered to move had their bedroom near the nurses' station so the nurse looked in. There the service user was in their usual attitude

during sleep – their head tilted back and snoring. The conversation between myself and nurse ran as follows –

"That looks very uncomfortable."
"I have asked them if they want to be moved and they said they were comfortable."
"I doubt that, you have to move them."
"But they're asleep."
"They cannot sleep like that, it is not comfortable."
"But it's how they want to sleep."
"Are you refusing to move them?"

At this I could see a looming confrontation so I went into the service user's room, woke them up and moved them up in then bed until they achieved the position that the nurse was happy with. Unsurprisingly the service user was not best pleased, shouting complaints about being woken up and thundering that they were going to report this to their relatives in the morning. They spent the long night sleepless and I was similarly unnerved about any complaint flying in about me; more than that though I was just totally worn down. I had attempted to satisfy this service user, seen, as I viewed it, to their needs and requirements and respected their wishes, yet if I had persisted in defending what had been their choice of sleeping position I would have found myself in trouble with the nurse.

As it turned out *I was in trouble with the nurse*. I was reported for refusing to follow instruction but only found out about it as the nurse had taken the time to tell the rest of the shift team the how, the why and the what that she was going to complain to management about.

It was perhaps this fact – an unsurprising one – that she had broadcast her intentions to all and sundry, that she had put her own spin on it and that she affectively was bragging about just how she was going to shaft me - that shattered the last vestiges of any hope of a working relationship.

I went home utterly depressed. It was the beginning of the end.

Yet I pondered the almost inevitable questions that I have never ceased to ask myself in the time since – although with the passage of time these have been based less on emotion – to believe that you are so disliked by someone for something, you know not what, apart from the very fact of being you, is a depressing and demoralising feeling – and more as a focussed post-mortem as to reasons or persuasions that lead to the collapse of trust and so as to draw lessons for the future.

Professionally the questions were easier to square than personally, this boiled down to the apparent conflict between being directed by a more senior staff member and being dictated to.

Could then, I have done something – anything - professionally differently?

Undoubtedly. I could have followed every edict, dictum, direction, and demand, however contradictory, however unnecessary; however much it may have conflicted with what I believed was the best possible care I could give to those I supported. Yet where would it have left me as a carer? A simple automaton drained of all instinct and perception and simply willing to follow orders? This, I believe, is where true madness lies in care – when a CW surrenders all thought and inspiration to arbitrary command. This is what produced the colossal Mid-Stafford Hospital goat-fornication – CW's and junior nurses surrendering their caring and nursing instincts, frozen from doing anything unless it had been handed down as edicts from up on high, bound by rules and prevented from employing not just their professional duty of care but, more crucially, and ultimately, their humanity and compassion.

I want to make clear *that in no single incidence did any one demand from this nurse openly and clearly put in jeopardy the safety or direct health of any one single service user*, but I believed then, as I still believe now, *these demands were at many times inimical to the comfort and wellbeing of service users in general*. Collectively though they combined to take away my own judgements based on having the most contact time with each service user, they also introduced different priorities that neither reflected nor responded to each service users particular requirements at any one time. These priorities were not based on any material facts effective at the time but were seemingly dictated by the nurse attempting to impose her will *not just on CW's and how they delivered care, but on the very care itself.* It was as if this nurse believed that she could in fact control every aspect of care *including service user's individual and changing requirements and needs as personally expressed through choice and respect of wishes.*

This was, and is, plainly ridiculous, but it was also dangerous in the aggregate – once a care team surrenders their own judgement, once their care is conditioned to only to act when told to act then you have care that is failing, and failing care ultimately leads to undue suffering and – as the Mid Staff's example shows – potentially premature death.

The episode with the service user who was prone to plunging BMs on a morning was indicative of this – I had, as a responsive carer, tried to find a solution to the very low morning BM levels and, through trial and error, had luckily hit upon a solution that least discomforted the service user and hopefully forestalled something potentially worse happening.

This was not because I was a paragon of care virtue but simply me, as a carer, doing my job and employing my caring instincts as they had been guided, honed, developed and passed on to me by other carers – in other words finding practical and workable ways to tackle pressing issues – *this is what care is as I understand it;* without this then I must have been dead wrong in all the years I spent in care in believing I was supposed to be a dynamic part of a care operation, not a passive one. I still believe that it is an essential part of care, good care at any rate, that Care Workers should be allowed and encouraged to employ their caring instincts in the drive to find the best solutions to the most pressing problems. Without this autonomy then carers might just as well be sentries, or slack jawed observers playing no contributory part in the health and wellbeing of those they support.

But this was only the most resolvable and least difficult part of the shoals that had brought this professional relationship to grief - what of the personal?

The constant questioning of everything I did, the separation of anyone else but me from blame for a particular incident or mishap, the confrontational attitude so readily adopted, the threats and the uttering of these to others and the apparent vindictiveness at the base of it was hard to fathom.

At the brink of paranoia when I was in the midst of this episode I asked, as the Philippino CW had done with me, for others to observe my interactions with the nurse in question, for them to decide for themselves if I was making more of it in my mind than was justified by the facts. Without exception 4 different CW's all found my treatment exceptional and personalised. Even more than this 2 CW's I had never worked with before and who *I had not asked to observe any interactions,* both commented that this nurse appeared to have, respectively

"A major downer on you."

And

"What's her problem with you?"

To the latter I felt like replying – I wish I knew.

And so nagging questions spilled up in my mind, two of which I could not avoid retuning to time and again.

The first was what would have been the reaction of others if our cultures had been reversed – if I had been an African CW and the nurse white British – would I have had grounds for at least making a case of racial discrimination? Could not a racial element at least have been admitted into the question? I do believe in this role reversal I would have been justified in feeling this, if not voicing my feelings. This notion was not groundless – this nurses reputation of "having issues" with South East Asian workers would only surely have lent some support to this behaviour as having a racial edge – after all, and again in our role reversal – if a white UK- national nurse had gained a reputation for discrimination prior to the episode with me then there would be grounds for further explanation.

Having said that I reject the full-on notion that the decline in this working relationship was based on race alone, instead I believe it had something to do with it but was more potently mixed with the second question that raised itself.

What, I couldn't help wondering, if our roles gender-wise had been reversed – in other words what if I was the only female on a shift team and the nurse had been male? Would I have had a case for sex discrimination? Certainly corporate cases of sexual discrimination in male dominated industries have been made on much less. It is hard for me to view this episode in any other light than my gender played some role in the hostility I ran up against.

For what it's worth I believe these two factors mixed – the latter dominating the former – but nevertheless combining to make myself the object of this nurses intense dislike. The irony of what I have covered before in my conclusions only added to their bitter edge.

I believe I was seen by this nurse in terms of colour and gender as a base for her pre-existing prejudice free from any empirical evidence – she looked at me and saw a white mature male and expected me to doubt and to challenge her authority, I was expected to be a troublemaker so she found evidence for it, she expected me to ignore her, and so she found evidence for it, she expected me to have a lack of respect for her, so she found evidence for it. As I mentioned in the above – if you expect something you can easily find facts to support your expectation – a subjective rather than an objective conclusion and one replicated in the more traditional forms of racism outlined above.

As it is all this mattered little, I was like a boxer hit by a clean punch in the first round who battles on gamely but knows that ultimately the fight is lost, that it is only a matter of time before another punch gets through and that this one will put him away for good.

Depressed and disillusioned I waited for the next haymaker to find its way through my tiring guard.

It duly did.

CHAPTER XX

Unpicked

The death of care comes invariably not swiftly but slowly and stealthy, for those who have experienced such a terminal point is seems like some predestined fate, appears as if all along this job death has been waiting for you, silently, crouching in the future patiently marking time until the distance between it and you closes to sufficient proximity for it to strike, for it reach out and clutch you, drag you in and swallow you down whole.

Dramatic as this may seem its truth is undoubted in my case. When the end came it felt almost to make sense in a twisted and ironic way, as if this was the only way it all could end, that there was no other way I would or could exit care, ineluctability it could be called – the lasting unavoidable drawdown into the grave.

The paralleled with the mortality that this description draws is not accidental nor overemphasised, with so much decay surrounding me, with so many people almost pointlessly expending their dwindling time right through to a point that seemed to have been have prepared for them ahead of time – running out front of them – that it was,and is, impossible to view the end of my own career in any other than such terms.

I witnessed in residential care – and as I have tried to explain in all the preceding - the way lives become so characterless and reduced that in the end it is difficult to see any individual existence at all that is not in some way whittled away to time marked on the slow march toward annihilation. As I have been at pains to point out this does need not be – nor should it be - the case – but it is.

So too with care. No matter what the good intentions upon starting out as a carer, sooner or later everyone becomes worn down to the point of blindly carrying out care "tasks" by rote, amalgamating individual people to the amorphous collective of "service users" and "residents" and becoming part of the problem that perhaps with the high ideals of inexperience they may have once thought to challenge and – this the ultimate act of hubris – to change.

My cynical tone here will leave any reader with little doubt that this is the mental place I ended up in; therefore if the slow draining of meaning precedes death in life – indeed makes mortality seem evermore so an immediately impending and therefore fully imminently expected eventuality, then so too in care; when I reached what marked the end of my care career it felt like this was the place I had always been heading, as if I should have expected no other fate, and I felt foolish too – I felt that I should have seen this coming all along.

And yet I was taken by surprise, shocked even; it was only later that it seemed to all make some kind of sense.

Is that what real death feels like I often have often wondered since, do we remain conscious enough after death to see our whole life in the round and the meaning of it that we missed in the living of it; is this what paradise or heaven or nirvana really is – seeing all things fall into place?

Looking back on my life in care the end of it then appeared wholly suitable, utterly predictable – but only so after the fact; approaching it from before the fact I was as unconscious of events as any one of the tens – perhaps even hundreds - of dementia sufferers I had supported over the years.

It was this total ignorance of the inevitable that made the final show down, the mad-silent sunset of my care career such a slow motion car crash that was as ridiculous as it was brain and gut-tearingly wounding. And it was this total ignorance that made me surprised by the fact that any thoughts I had that I worked in a care system that was rooted in providing unified care outcomes - supported by my fellow CA's and clinical staff - that improved and enhanced the lives of those we were there to support and care for, was a complete, total and even ill-disguised illusion.

It also liberated me in the most brutal fashion of the vast stupidity I had duped myself into thinking - that care was a place where free thinking and initiative were looked on as something good rather than something subversive; finally and ultimately it comprehensively destroyed any aspirations I had that I could make any substantial difference in the lives, and the quality of them, of those I supported and cared for.

It is also a cautionary tale for any CW of how not to endear yourself to clinical staff or to show too much forward thinking or intelligence to those in more senior positions.

The single most important fact I drew from the end of time in caring – and a depressing fact at that – is that the ideal carer for service operators is one that follows instructions to the letter without challenging or questioning the basis for them, who acts against any caring instincts they may possess and will follow all with the conformity of a cowed individual labouring under a totalitarian regime.

The NHS is often portrayed as a Stalinist organisation that bends the knee to centrally imposed directives, however arbitrary and self-defeating, in which an individual can be destroyed by the machinations of a "party machine" that demands obedience in thought and action and who's whims give the impression of discipline as being conducted according to Beria's maxim – you bring me the man, I'll find you the crime.

In the private sector though - although such an apparent dictatorial regime may not exist due to the fractured structure of the profusion of care providers and the decentralised nature of authority in residential homes - instead a form of Maoist control is alive and well - where home managers, senior nurses or even Senior Carers can contrive denouncements that would not be out of place in cultural revolutionary China. While it may be possible to be ground under the wheels of bureaucracy in the NHS, in the private sector you can have your career and livelihood – and your very sense of self - destroyed by one or two individuals with a grudge.

Perhaps this will strike as an exaggeration verging on paranoia, but the last dog ends of my care career will hopefully provide a corrective to that impression - and show just how cruel and savage the business of care is.

The genesis of my downfall was as comedic as it was heightened with tragic irony. No black comedic farce could do it justice, the only restraint of hilarity being the emotional cost to me and the revelation that I worked in a Care Less business. And to think it all started with a cock-end.

Turning up for a nightshift one evening I arrived early to find one of the Senior Carers writing out the handover report – or rather trying and failing to write it - their forehead was cradled between their thumb and fingers and in their other hand their pen hovered over the paper, occasionally it made almost reflexive, jerky attempts to apply it to writing only to withdraw at the last moment. I clocked in and was about to walk down to the staff room to get changed when they called me back –

"I don't know how to put this." They said

"Put what?" I asked pulling on my uniform top. The Senior then went on to describe the horribly painful sounding ailment of a catheterised service user – they had developed a blister on the end of their penis.

"Right at the end?"

"Oh yeah, right on the Bobby's Helmet."

He said using the widely known slang description of the end of the male member

"Ouch. Why are you struggling how to put it?"

"Well I can't put Bobbies Helmet can I? I can't put bell-end either really."

"I suppose not."

"So how should I put it?"

I ventured toward an area I knew was tricky, and I speak literally as well as euphemistically, as I hesitantly and guardedly put forward the medical anatomical term for the area that the senior was describing.

"You could put the glans," I said "this is what it's called, anatomically speaking, but if you do please don't say you heard it from me."

My caution here was not without cause. In care I had always been acutely aware that using medical or anatomical terms in front of nurses was a big no-no. Nurses, I had learned, were particularly sensitive to any CW venturing beyond the very limited bounds of their "expertise" – which they viewed as wiping bottoms, feeding residents and not killing anyone either deliberately or inadvertently. They jealously guarded what they saw as their unique domain and would be ill disposed to anyone who crossed the line onto what they perceived as their turf.

A classic example of this had hit home to me a few weeks before. A service user who was incontinent of urine had a "habit" of being dry at every check through the night or day then suddenly become soaking very shortly before the shift change at 8am/pm - so causing all kinds of friction between night and day shift - as the night shift/day shift would check that they were dry at the latest possible moment only to learn on coming in for their next shift that the service user had been found "soaking wet" thus levelling the accusation that the night/day shift had not monitored this service user properly. One of the night staff CW's - who was the main carer on the floor where this service user was roomed – had suggested during the handover that the service user in question appeared to be –

"Retaining urine" only to "let go" in the morning.

A World War One flying ace could not have shot down an enemy aircraft as fast as this CW was taken down by the two nurses present at the time.

"They're not in urine retention." One shot back, the other sniped rhetorically and sarcastically – "And what would you know about urine retention anyway?"

Of course what the CW was meaning was that this service user appeared to be - unconsciously - holding onto their urine over the day or night time only to let it go all at once. This in itself was plausible, I have known many "incontinent" service users still retain the vestiges of continence that the average adult has, only for them be suddenly overtaken by the impulsion to urinate all at once. The CW's big mistake though was to use the term "retaining" which both nurses had leapt upon.

Urine retention, or, in medical terms *ischuria*, means something quite different –

"A lack of ability to urinate."
http://en.wikipedia.org/wiki/Urinary_retention

and is a symptom of a deeper pathological or physiological condition, something quite different from someone holding their water through the night or day. To any reasonable and less professionally insecure individual or group this inadvertent misuse of a more complexly implicit term by the CW would normally have been understood as such and disregarded or glossed over, and, at the very least, the need to embarrass a less medically trained individual would have proved wholly unnecessary.

Or instead, in order to avoid future misunderstandings through the misapplication of a medical term – with the attendant risk for individuals through misdiagnosis - perhaps a polite and finessed reply that this service user, while they could be "holding" their urine, was not retaining - as this meant something very different - would have been appropriate. However it was, and is, symptomatic of a kind of nursing neurosis that persists in care that this misspeaking was used to slap down the unfortunate CW in order to prove how much they – the nurses – knew, and how little the CW - both relatively and absolutely - did.

I was there in that handover and felt acutely embarrassed myself, but wasn't quite sure though who I was most embarrassed for – the CW who had been humiliated – or the nurses for being so incredibly arrogant and insecure. Either way it was a reminder to me to steer well clear of *any* medical term at all.

So when I ventured to put forward a term I knew to be right but for which I may not necessarily be thanked for introducing I placed the – "you didn't hear it from me" - caveat on it.

I should have known better.

The handover started and the Senior came to give the handover report; when they came to the service user with the painful blister condition they reported in a loud and penetrating voice –

"Mr X has a blister on his..." then half turning pointedly to me *"glans,* that, according to [my name] is what the bell-end is called."

I started to shrink back in my seat, but the worst was yet to come, the Senior turned back to the senior nurse coming off duty *who was also the deputy manager* - and the Most Unpleasant Person In Care - and shot away –

"Is that right? Is it called the glans?"

By this time I was almost under the table and felt myself turning the colour of the place where the offending blister was located.

The senior nurse/deputy manager looked blankly back for a moment. This nurse was a mental health nurse but at the time I was under the impression they were general trained and either way I was sure that they knew full well what the tip of a penis was called. I waited for the expected affronted intercontinental ballistic professionally loaded response to be launched back at me; instead an uncomprehending look stared back at me, then switched back to the Senior, there was silence for a moment. The Senior carer waded on –

"Is that right?"

The senior nurse mumbled back,

"I'dunno."

Suddenly I caught a flash – in all honesty I don't know if this was actual or if I just imagined it in retrospect, yet I'm sure it was there - a sudden sharp-eyed snipers shot of a look right down the barrel at me. For an instant I thought – that look doesn't look right – it was inquisitorial and almost accusatory, but it was only there for a flash, then gone.

Right then and there I knew one of two things had happened. Either I had struck on a term, unbelievably, that a senior nurse was unaware of – very bad for me – or, and worse, they knew full well the anatomical term but had read the situation as a set-up by me in order to try to catch them out – impossibly, incredibly and inconceivably bad for me.

Silence came down for a moment after that flash/stare before the nurse coming on duty moved the handover on.

I knew right then and there, one way or another I was shafted, truly shafted, madly badly totally as I noted in both the last chapter and the head of this one the very worst thing a carer, *any carer,* can do is display anything resembling intelligence, independent thinking, a knowledge over and beyond what they are *supposed* to have and any sign that they have a capacity to learn or to grow - clinically, psychologically or professionally. Yet I would go still further – it is career suicide to be anything other than an obedient and rigorously thought free "work unit" - to work guided by nothing else but the simple and specific instructions of those further up the chain of command. And I use the term "chain of command" with good reason for, like in any military setting, the chain of command is sacrosanct, regardless of the fact that in contact with the real world, in contact with the basic common human level, in contact with anything resembling compassion – these orders may be ridiculous and in fact contrary to the good and wellbeing of those who's care we are charged with – the very fact that these orders are issued down through this chain of command though gives them an authority that trumps all others. To default from this chain of command, to deviate from it, to display that you may possibly have ideas or instincts that identify yourself as standing out, is a professionally capital crime for it means that the "work unit" may possibly, one day, God forbid, go from thinking to the next step – asking questions – and then – Christ almighty, they may just ask the multimillion pound questions that expose the very edifice that modern care is built upon – a senseless process designed to streamline the throughput of biological ATM's – service users – with the least friction - clinically, psychologically, mentally and physically - possible.

This is why intelligence or evidence of knowledge is so quickly pounced upon and beaten out of the system – because to stop the process along the lines of independent thought is to forestall the exposure of the banal impersonality of care.

I knew this back then – perhaps not quite in the visceral way I know it now – but along the lines of a basic knowledge that something is dangerous territory – like knowing that a minefield is dangerous to walk into without having to be shown *how dangerous* by someone getting their legs blown off.

Therefore I knew I had blundered into that minefield and the fact that, over the next few weeks, I still had all my limbs meant nothing apart from the fuse was still down.

No one will have no idea though just how much mental turmoil I was in during this period when "nothing happened". In my mind, nearly all the time when it wasn't occupied with a pressing matter of the moment, I went back to that handover, over and over I could see the senior nurse's/deputy manager's face as the Senior had uttered the word *glans*, had followed it up with that question that had elicited that blank response. Time and again that face haunted me because in that face there had been the blank look but also the frantic movement of the eyes that had performed a myriad of movements and nuance during the exchange. I was convinced – or rather tried to convince myself - that I was making it all up in my head, that I was imagining things that hadn't actually happened and yet was "making" them happen in replay; and yet, and yet, and yet I was *sure,* sure as I could possibly be that in that vignette of eye movement held the total criticality of the event, held the whole story, only a fraction of which had leaked out into words.

In my mind I saw first the lazy focus of the senior nurse's/deputy manager's eye on the Senior CW as they started making the report, then saw them rapidly come into sharp focus on the utterance of the word *glans* then hardened into a gimlet stare at me as that word was attributed to my mouth and originating from my brain. The blank look was still there in my mind but the eyes held themselves steady, staring at me without trying to be so obvious as to be staring at me.

"I'dunno." - what was actually spoken was almost insignificant next to the eyes that found fixed and meant to finish me.

To those outside care waiting for a shoe to drop is the worst of all moments for those working inside care, worse than even the consequences that you just know are coming; everything you do seems to be in suspension, you look at each and every action, each and every reaction of each and every single service user to see if anything can be "read" into it – will this look good, will this look bad, how will this play, should I have said this, should I have done that - all of it, very single scrap *of being* seems to be there up on the block. This is stuff that nervous breakdowns are made. My own nerves felt like they were a shredded cat's cradle; still, no matter how insane you feel you sense that something worse is coming down the rails, or, as Shakespeare put it – "The worst is not So long as we can say 'This is the worst."

Then, one morning, after a night shift I was summoned to the manager's office. On the way down there I knew "the something" had arrived, I walked down the stairs like I was going to the gallows, I was cold sweating and my spine felt like it had been removed plunged in liquid nitrogen then stitched badly back in.

The manager – who I'm not sure was actually busy or just pretended to be – told me to go and wait in their office, whether this was part of the "softening up" process I wasn't sure – if it was it was dreadfully effective. I remained standing in the managers office as I thought if I sat down I might not be able to stop my knees from knocking together; the day shift CW's passed back and forth past the open door and I got the impression that rather than just passing in the course of work they were actually taking turns to come past and stare in at me, like I was some biological curiosity in a jar, or else they were hoping to catch something of why I was standing there, or maybe they were gaining some vicarious thrill in seeing someone on the steps of the gallows – all of this, it was turn out, proved to be true, it turned out that everyone – and I mean everyone who worked in the building - knew why I was there already, well before I did.

The manager eventually entered with the night nurse and closed the door, I was invited to sit down while they gathered some paper from the desk, then they came and sat down across the low coffee table from me with the night nurse hovering in attention.

I would like to recall what was said, what words were actually used but the truth is after the opening sentence I hardly retained anything was what was said – or rather I knew *what* was said but not *how* it was said.

As far as I can recall it went along the lines of -

"I have received a number of statements detailing complaints about your work, including that you are very rough with residents, you don't interact at all with residents but just enter their rooms and do what you do without explaining anything to them, you ignore resident's requests and your general attitude is one of confrontation."

I think it went something like that anyway, I couldn't really tell, as by this time I was feeling sick, that swelling deep in your stomach that is the foreshadowing of vomiting, my head felt like a helium filled bladder on a stick and my eyes had trouble focussing, the tongue in my mouth was sandpaper in a desert so it was no wonder I couldn't retain the form of words used. I recall I stammered out something like –

"When?"

My voice, I was aware, sounded thin and reedy, dry and high pitched, like schoolboys – very apt because I didn't just feel like I was back at primary school getting a carpeting, *I was back at primary school*, all I needed was shorter trousers and less crow's-feet – I already felt a good 2 foot shorter and shrinking all the time, shrinking back to where memory was mixed unpleasantly with the present and I could almost smell the washed out disinfectant that the cleaners used to use in schools to mop the halls when everyone had gone home and only those "kept behind" remained.

"On several occasions." I was told bluntly.

"But how? What were the instances?" incoherently I wanted to hear something more concrete, an example, an actual event that would describe in more detail what I had done wrong rather than the just these vague but viciously stinging accusations.

"I can't tell you that."

"Why?"

"Its confidential." Perplexity set in, confused and disorientated my mind scrambled after hard facts.

"Who's made the complaints?" I clutched at straws;

"I can't tell you that, it's confidential (how I laughed, hollow as an untolled cracked cast in bell at that one later) all I can say is that they all came from your fellow works who you work with every shift and are 4 in total."

At least that was the drift of what was said.

You could say the situation was Kafkaesque but in the fullest extent; like the unfortunate Josef K my crimes consisted not just of a professional rebuke but also a moral and existential one.

"Am I getting disciplined? Sacked?"

I don't know why I asked that, the lack of anything other more suitable to say I think. Yet already I recall my mind scrambling, trying to think of incidents or actions or anything that may have produced this turn of events. Again the following exchange went along the lines of -

"Not at present, however a formal outside investigation leading to possible dismissal has been discussed at the highest levels, however it has instead been decided that from now on your work will be under close monitoring and supervision – I shall ask the nurse on shift with you to submit reports of your conduct throughout any shift - and if I get any more complaints then I promise you, you will be investigated, sacked and placed on the SOVA (Safeguarding Of Vulnerable Adults – now folded into the DBS check) register which means you will never work in care again, as I, and the company, take these complaints very seriously. Now I want you to take a few days off and reflect upon your work and how you do it and come back with a new and better attitude."

"Is this a suspension?"

"No, I'm just giving you a few days of to think about what you are doing and how you are doing it."

"I just don't recognise any of this. I've never treated or handled or spoken to the residents like I'm said I'm supposed to have."

"You may not but I have four statements to say otherwise. One, maybe two and I could have put that down to personalities, but when the 3rd and 4th came through soon after it raised serious questions about your conduct. Now I want you, like I say, to go away and think on how you do your work and come back with a better awareness of your situation, which is very precarious at present, and start to do your work *with much more respect and care for our residents."*

Now this last line I do remember exactly because of the severity of the pain implicit in it, just the same as the number 4, FOUR complaints.

What the hell?

I felt my mind reeling under the assault of such accusations, so much so I felt disorientated and almost physically battered, in fact a physical battering would have been much preferable to how bruised and kicked around I felt mentally and emotionally.

For a moment I thought I was going to pass out, then I pulled my stuff together and headed for the door which the manager was holding open for me. As I walked out onto the ground floor no one spoke to me, I was aware of other CW's – people I knew and had worked with - shooting fleeting glances at me then hastily looking away, one or two actually, literally, turned their backs on me. I clocked out and punched out the pass code on the door and walked to the nearby shop in a daze.

One thing ought to be made clear for those not familiar with care at this point. A statement is a very serious in care; they are actually more serious than complaints.

Complaints are often made by care staff either about each other or about nursing staff or even about management, these can be verbal or written and can also be considered the grit in the gears of any "normal" workplace; as such care is not unusual; anywhere where people work together there will be conflicts and disagreements that need man management of one sort or another, so complaints are usually handled informally and swiftly.

A statement though is something of an entirely different character and order, in care statements are nearly always deadly, and sometimes fatal for a carer and their career.

A statement is a formal outlining of either a specific or culminative act(s) or behaviour(s) that are used when serious events arise. As they are written they become part of the written records on which care stands or falls, and they can be used as a pretext and sometimes as the foundation for a serious disciplinary, up to and including sacking.

Infinitely worse though than losing your job such statements, if stood behind by a manager, a nurse or a carer can be used to refer an individual to the local safeguarding board, if these statements and the managements backing of them are substantive and reaffirmed it can lead to a carer being placed on the barring list held by the Disclosure and Barring Service – just as I had been warned. That means that potentially such an individual can never work in care again. Even worse than this though is the scope of activities for which a DBS clearance is necessary – any job with any possible or potential – however tenuous – contact with any vulnerable group – means you needs a DSB clearance. Toilet cleaner? DBS. Caretaker? DBS.

Statements too are "sticky" - once they have entered the chain of paperwork in care there they stay to be held indefinitely, so if you are to run into trouble again they can come back like Banquo's ghost to condemn you in tandem with any further statements made, they are the care equivalent to the Sword of Damocles.

This of course is effectively constructive dismissal or dismissal on the interest free credit plan – in any other work sphere this would put any employer or group of employers on shaky ground for bringing disciplinary's or dismissals, in care though different rules apply, once statements are on file you are walking the line every time you come into work, every action, every interaction, every word can tip the balance, end up in another statement and by culmination they can destroy you.

Just to re-emphasise – on the basis of statements and following only the most cursory investigation – it can be possible to find yourself locked out of care inside of a month – so that's not just your job gone but any career in care.......period.

I don't know how long it took me to walk the short distance home. I remember going to the shop and buying some pop, then I remember smoking slumped against some wall. I must have meandered somewhat drastically because I found myself in unfamiliar streets – yet the way I was feeling all streets felt unfamiliar to me – a strangeness descended upon me that made me feel quite different from what and who I was, it felt like I didn't recognise myself, that I had become, in the words of poet Rimbaud – another.

I do recall finding myself, totally accidentally, back outside my place of work but how I got there I don't know – I must have travelled in a big loop like someone with one foot nailed to the floor. I thought of going back in there to argue my case but thought better of it and set off for home......again.

The reasons for my twisted perambulations were many, as many as everything that that short carpeting had raised, but 2 agonising thoughts kept tumbling over and over.

Firstly, and most painfully, there was that acutely lancing parting shot to –

"Go away and think on how you do your work and come back with a better awareness of your situation, which is very precarious at present, and start to do your work *with much more respect and care for our residents.*"

Much more respect and care for our residents? – How painful that was, implicit in it was that I had been mistreating them, no, let's go straight to the nub of it – abusing them. Of course the manager had steered clear of the actual "A" word, been very careful to, but that was only because to have mentioned it would have been to trigger an investigation, yet in all they had said the "A" word was implicit.

This sickened me more than anything; to think that I could be thought capable of being an abuser, of being seen as one, as being virtually labelled as one. Sure it wasn't alleged intended abuse but that generated by indifference and lack of compassion or empathy. Somehow this was worse – to set out to do something bad you know what you're doing, it is premeditated, and so the consequences expected; but to do a thing unconsciously, to *lack care,* to be, regarding very vulnerable people, *care-less* – that raised the fundamental question of my suitability to be doing what I doing; more than that it put my whole being into question – the person that I was, or at least the person I liked to think I was.

Yet in it all I could not square this with anything I had done. The allegations appeared to me as if they were all directed at some other person, *not me,* I just could not contemplate them as a reality, as something I *owned* - as somebody - or rather, *as something - I was.*

This was the core of finding myself feeling as "another" as someone different from myself.

Rollockings at work are never pleasant no matter where you work or what you do, but in care they have a much fiercer edge – perhaps no other work sphere links work so much to your sense of being – and that makes care work as much an ontological as much a remunerative occupation – in care you put your very soul on the line - *how* you care, or *how you do not care* goes to the very centre of who and what you are.

To be bulled for not doing your job properly in care means, by definition, you are letting down the people you are there to care for; you are failing them at the most basic of levels of humanity. In this case I was accused of treating them badly, serving them poorly and degrading their existence. This, more than anything, made me feel sicker than any other element implicit in the ferocious criticism – first - the thought I could be viewed as being indifferent and unfeeling - was wounding, but second, and much more painful, was the thought that *maybe I was.* Had I been fooling myself all this time, had I been lying to myself that I was actually a good carer, that I was suited to care, that I had the skills or *was the person* to care; was I – ultimately - *not the person I thought myself to be?*

This horrible question tore away at me like hounds falling on a grounded fox. Back in the emptiness and silence of my house – when I eventually reached there - the savaging I felt morphed into an oppressive internal pressure, it felt like my skull was not large enough to contain all the thoughts, the terrible thoughts that pushed and pressed at the crown, at the temples, at the back and down into my spine. If I wasn't who I thought I was, was not the person I believed I was, then who was I?

Yet once again I came back to the second nagging thought that haunted me - the thought that I could not think of one incident or episode that could justify this criticism, I could not recognise myself or my work conduct in what had been so woundingly described and ascribed to me. If only I could have thought back on *even one incident* of improper conduct perhaps I could have come to terms with the accusations made against me - to *own* them - but I could not. Yet there they were, in ink and paper somewhere in the manager's office, there they were black and white denouncements of who and what I was.

The truth in care is that it doesn't matter actually if you are a good carer or not, it's not *what you are* or *what you do* but what you are *alleged to be and do.*

Over the next few days off work I dedicated myself to just what the manager had advised – I reflected on my work and more – examined my conscience; as I did the more sure I was that these allegations were groundless. I spent literally hours racking my brains, ransacking my memory for any clue that I might be *who and what I was alleged to be*. In all this I can truthfully say I could not come up with a single instance to justify the statements, to give me pause. Yes there were times I had not been as attentive and "present" in my work as I could have been, but nothing like indifference, thoughtless action, mishandling; there was nothing that I could rebuke myself for even though I tortured myself over the smallest detail to see if I was only indulging in self-justification.

Still my conscience wasn't clear – why? Because even though I was as certain as I could be that I had done nothing wrong there were still those statements, standing over and against me in silent, powerful, nigh on irrefutable, admonition.

Day after day, night after night through those days off I either lay awake or sat in a chair paralysed by these thoughts, sucked under by them, they rose up to punish me over and over. The gut twisting that attended them was as bad - a grinding deep down in my stomach, the horrible ratcheting up of so much tension and self-doubt – so much so in fact that I developed stomach cramps every day, more than once I found myself on all fours arching my body up to see if I could get rid of them. Nothing worked.

Eventually I surrendered and went to the doctors and came away with a prescription which at least did away with the cramps but not the thoughts.

The fact that could not reconcile the complaints with anything I had done only led me back to the nagging question of *who* had made them, and *why* they had made them – if I could not accept what was said about me, if I thought deep down – as I did, and still do – that the accusations were false and baseless – then why would people – people I knew and worked with - knowingly make false statements about me, in the full awareness of just how damaging they would be.

This was, as I have already noted, like a denunciation in the culturally revolutionary China – me, who I thought was liked and in some small degree respected by my colleagues had found myself on the absolute defensive against vague yet damning allegations and reduced to the lowest of the low, like a party official banished to work the fields with a trowel and his bare hands, constantly under suspicion, regarded as "unsound" unreliable and alienated. Even though I tried to focus on the allegations my mind kept turning back to the who; who had done this to me?

Over the next few days and through a process of elimination I narrowed the field down – I knew one source - a lazy CW who had resented the fact that I asked them to give me hand with a resident and so stopped them from partaking in their usual work – slumped in front of the TV in the lounge and doing as little as possible – because they had bragged to all and sundry that they had made a statement against me, because, as they were related to me as saying –

"He can't tell me what to do."

Then there was another worker who was only ever happy if they were causing trouble, they thrived on it and looked at it as a way of generating more gossip, which was their favourite pastime. Then there was another – a pleasant enough go-along-to-get-along individual who I knew never let anything get in the way of their own self-interest; they were master manipulators of emotion in others, capable of twisting the truth as well as other people to suit their purposes.

This one hurt more than the others, who I could dismiss in one way or another. This carer I had always tried to help, to look out for, to defend and back up. It hurt me to think that I could have been shafted out of pure self-interest, yet here I could see an angle – a way of endearing themselves to management by turning a colleague over.

But the last - here was a real dilemma as the only possible person left was someone who I had worked alongside for 2 years, who I trusted, who I had once covered for in a similarly baseless disciplinary and in doing so put myself on the line to save them from getting the sack over a far more serious complaint made by a residents family (this did not involve abuse but the use of bad language in the presence of a service user), someone who I thought of as being as much of a friend as it is possible have in care.

Had they done this too?

The problem was there *was* no one else, it couldn't *be* anyone else.

This almost destroyed me.

As I have said before, care works on trust – you have to trust the people you work with because without that trust everything falls apart - you need people to help you, to vouch for you, and to work together with you. While you do not expect them to cover-up anything – I would never expect that of anyone – indeed I had always told the people I worked with to tell me if my approach, if the way I worked, could be improved – you do though trust them to come to you if they have a problem with you or your work, and to tackle such problems collegiately. If this person really thought what they had apparently put in a statement about me and my work, then they could easily have raised this problem with me one on one instead; I would have listened, I would have paid attention on the grounds of the seriousness of their issues with me and the fact that (or so I thought) they knew me well and I knew them.

So how had it come to this?

Why?

It felt like a horrible betrayal.

That first morning, on my long walk home I decided to tender my resignation. After all, how could I work there now? How could I trust *anyone* I worked with now? How could I live or work with the possibility that at any moment in the future I could be subject to more blindsiding statements that would do for me professionally?

Yet when I thought about it I saw what a catch-22 I was in – if I tendered my resignation now it would look like I was guilty as charged, as if I was jumping ship before I was pushed; if I stayed I ran the gauntlet of being buried in more statements.

The more I thought about the allegations though, the less substantive they appeared. I spoke to a friend much more experienced in care than I about statements made against me and who I trusted to tell me the truth as they saw it – they pointed out the fact that in over 2 years working for the company I was employed by I had never had a single complaint made about me – had I suddenly changed the way I worked?

No.

Had I done anything that was, or could possibly be construed as "out of bounds" in the normal run of caring?

No; and I had scoured every action I could call to mind over the last weeks and months in extremis.

Well then if you didn't have any problems before now with the way you worked, *this problem* seems to lie elsewhere, they counselled.

They had a point, nothing I had done before now had been deemed blameworthy; why then suddenly should that cease to be the case?

This hinted at something darker going on, something behind the scenes.

The more pressing issue at that time and over those next few bleak days was to make a decision about my immediate future, should I stay or should I leave? I had to weigh the thought that to resign would confirm all the worst suspicions surrounding me against the very real possibility that I could be subject to death by a thousand cuts if I stayed.

Had I known what I was to find out a couple of days later the decision would have been even more pointed and double edged, as it was I was blessed or cursed with ignorance of the situation in its entirety and took the decision to stay.

For better or worse I felt the impending need to somehow "clear" my name, to expunge the allegations against me, to show I wasn't who or what I was painted to be.

While this decision was based as much in regard to reputational risk, self-respect and an amount of bloody mindedness, it was also mixed with a heavy dose of self-criticism. Even though I felt all the allegations against me were unfair or misrepresentative - or even malicious - I took the words of the manager in hand practically – I consciously mentally redoubled myself to scrutinise every bit work I did and how I did it, I would go back to basics and be more aware of what I was doing. I knew I wasn't perfect but maybe I had let my eye slip off the ball, I intended that that shouldn't happen again.

When I returned to work my decision not to resign seemed hasty and, to a degree, ill-judged; it confirmed all the worst fears I had about my reputation being dragged through the mud – it turned out everyone in the building knew not just that I had been taken into the manager's office – that much had been made obvious – but what I could not have known or counted on was that the whole content and substance of the conversation that I thought was private and confidential was in fact common knowledge throughout the building. In fact the compounding nature of gossip had elevated the whole incident way beyond what was already an horrific and humiliating fact.

It turned out that lurid breathless talk had me being "carpeted" by the manger not once but 3 times for 3 different "offences" (where the 3 times had come from I don't know, unless my doppelgänger had been knocking about and had been dragged in there a couple of times more that I didn't know about) that I had been warned about "assaulting" residents, that I had in fact injured residents and that I was on my final warning - none of which was – obviously - true.

While these rumours took time filtering back to me what was immediate was the reaction of all the staff in regards to me, there was almost a collective rejection or at least a septic reaction to me.

In care if you are subject to a disciplinary or even, as I was, a serious "talking to" it is as if you have contracted a contagious disease, people keep their distance, look at you sideways and have a curt and lopped off way of speaking that is usually reserved for bureaucratic (un)civil servants, there is a distancing and an alienation of you, you are pushed to the periphery of whatever social atmosphere exists in care. You suddenly become intimately aware of, when it goes to the mat, how alone you really are in care; no one wants to be associated with a care-leper - as you feel yourself to be - because you have been marked, stained and pointed out, you are rejected - all I was missing was a bell and cowl to cover my leprosic contagion.

In a lifetime punctuated by depression with no apparent cause never have I felt such a pit of despair – yes - despair is not too strong a term for it - so linked to a single event as this one. Without being over-dramatic thoughts of suicide occupied more than a little of my time at work – the futility of going forward with a process seemingly designed to sap your spirit and soul was acute while the option of throwing in the job was made ever harder by the reactions of other staff – the seeming willingness of *all* of them to believe the worst of me, even though almost all of them had, at one time or another, worked with me and had never had cause to doubt my caring attitude.

It was like a frost setting on me, a cold shunning. Even people I regarded as close work colleagues suddenly found something *really important* to do when I walked into their presence, or they found excuses to do something else if I need help with anything. I found myself left alone on units where I worked as anyone paired with me discovered a hitherto unknown appetite to "file paperwork" whenever there wasn't a care task that needed doing. The nurse in charge eventually (and rightly as will be shown below) posted me to a unit where I was the only carer so that no one would have to work with me. I suppose the overriding notion in the minds of even carers who knew me well was that if I was on the verge of being truly stuffed they wanted to be nowhere in vicinity at that defining moment lest they get dragged backwards into it.

This isolation, in retrospect, was just what I needed. First of all it saved me from two really awkward situations. 2 of those who made Statements against me – the go-along-to-get-along carer and the carer I previously thought of a close colleague – the two who had wounded me the most – were on the same shift pattern as me. When I turned in for my first day back I walked into the handover room to be treated to the classic situation of the socially-excluded – a hushed talk between these two carers suddenly ceased as I walked into the room, one of them disappeared "to the toilet" – even though I could hear them talking loudly seconds later to someone in the corridor outside - while the other bent their head and dedicated themselves to fiddling with their phone - the definitive act of modern social distancing and disengagement –until eventually other shift members turned up to break the awkward moment. I felt sick and for more than a passing moment I considered just turning round, heading home and giving up.

The shifts where I worked with either of these two carers were excruciating; long silences – and over a 12 hour shift long silences mean *hours* – were punctuated by their absenting themselves totally from the unit we were working in, so much so I was reduced to calling them on the phone to ask them to help me to actually care for any service users. During such care the process was agonising; neither of us actually wanted to touch the service user, each of us deferring to the other for their own reasons – mine because my confidence had been shattered and I was no longer sure of what actions were deemed – or could be spun – as inappropriate or otherwise - so wanted the other CA to take the lead – theirs because they didn't want to get caught up in any action that they may find "useful" in the future to use against me.

The result was that hardly anyone on the unit where we worked were being cared for adequately due to this breakdown in trust; it became so obvious to the nurse that I could not work with either of these two carers effectively that the only solution was to move me. Besides saving me from this dysfunctional situation the lone posting was a boon *because* of the fact no one had to work with me, it meant that whatever I did was all my own work, I was not terrified that any move by me could possibly be construed or spun in a negative light. I came to treasure this time as it slowly returned to me some of the confidence I had lost. I saw that I was not a bad carer, not some kind of unfeeling monster, not a dehumanised and de-humanising presence. Alone for 12 hours with 15 residents I began to build a rapport that is not always possible when working with another CW, several of the residents – normally uncommunicative and reticent - became more alive with me; they talked, I listened, and bit by bit we built relationships that I had missed before. I also had the chance to reassess my work both before and since the carpeting I had received.

Over the next few weeks some of the awkward situations and mephitic atmosphere surrounding them resolved themselves. One of the carers that had made a statement about me – the go-along-to get-along-type – went on the sick for unspecified reasons – only later becoming apparent; while the colleague I had once trusted moved to another home within the company.

If I had been seared by this experience though I was soon treated to the transient nature of how critical situations can blow up out of nowhere and then disappear just as fast.

In my decision to stay I had submitted a letter to the manager outlining my position - that while I did not recognise the nature of the statements made against me that I would look hard at my way of working and would welcome any monitoring of it.

As a result of this monitoring I would seek a meeting with her in the next 4 to 6 weeks to check on my progress and to hear any suggestions or directions fed back to the manager by other staff or the nurse in charge of the shift and then to act upon them.

Time passed without further incident and after 6 weeks I asked for a meeting with the manager - the context of which was as follows.

Finding them available after one night shift I asked either to speak to her now or make an appointment to see her at her convenience.

"What about?" She asked as if my very presence – I was still trapped in the doorway not having been invited in nor told to disappear until a more appropriate moment - was indescribably tiresome to her. Looking at me squarely, as she did, like a corner brick in a blunt wall, she held all the appearance of someone who was being held up and prevented from doing something much more important even though she was occupied with nothing at that moment nor being diverted from any task.

"About my progress and the results of the monitoring and any improvements to make in my work."

"I haven't heard any more complaints or had any more statements about you if that's what you mean."

"Good......good." I morsed out, staggering over the code that this rapidly going nowhere conversation appeared to be couched in.

"Is there anything else I need to be doing?" I asked, struggling to make headway against this seawall of indifference.

"If there was you would know about it by now."

This last sentence seemed to be good news – I was in no further trouble – but was phrased and intoned in such a way that it felt like an admonishment.

"Is there anything else?" she said, again putting it in such a way that I felt like I was transgressing or intruding.

"No it was just in the letter I said I would meet with you to........."

"Yes I read it there's nothing to say."

"Ok.....er.......thanks."

At this she waved me away with a flap of the hand that was both imperious and casually dismissive. I felt like an ant being brushed off from an impeccably white jacket of an old colonial administrator.

I slouched off completely baffled and almost in as much of a spin as after that previous, horrendously crippling, meeting.

Less than 2 months ago I had been made to feel like a criminal, that I was in serious, career threatening, trouble and that I was under the most intense suspicion and scrutiny; I had felt that all the allegations made against me were taken so seriously that in aggregate they meant I was about one step from severe disciplinary sanction or the bullet. Suddenly now it seemed like I was inconveniencing this same manager over the same subject that 6 weeks previously had been dealt with in the most censorious of manners; she appeared to find something that had been such a serious issue back then as now being incredibly tiresome and bothersome.

This was my introduction to how one moment in care you can be almost out the door and in the next returned to the ant heap; it also, as if any further proof were needed, displayed how little CW's were actually thought of by managers; infinitely expendable one minute treated with cold indifference the next. If the issue had been so serious one moment (and, with the passage of time, it appeared to be almost literally "one moment") why was it now the subject of not even passing interest? The answer in my particular case was both more complex and symptomatic of the situation of all CW's in care. The complexity of my case I will come to shortly but perhaps the symptomatic element is actually more profound and has more resonance for the private care sector in general.

The fact is that CW's in the care sector are not treated with anything approaching a professional manner, and this had wider implications for care. To my naive self I had thought of my carpeting as part of a process – however unfair I thought the allegations to have been I thought that as they had been treated so seriously that monitoring and follow-up would not just be a consequence of the allegations but a necessity – an integral part of dealing with what they saw as a failing carer. I thought the process would be something like –

First, make a carer (me in this case) aware of issues raised by any Statements made against them,

Second - address the issues raised in those statements with the carer,

Third – that the carer, having had their attention drawn to these issues, and appropriately warned regarding them, a structure would be put in place to monitor their future conduct and implement an on-going process of review in order to see that the changes demanded had been absorbed and put into practice.

When I asked for the meeting with the manager I assumed it would be as formal as the one where I had been so viscerally criticised, that the manager would have asked for, and had available to them, information regarding how I was working, any further issues raised – however trifling – and an action plan for another period of review; and, after even this, a longer term plan so that I could keep improving beyond the merely adequate and un-troublesome to in fact become better than my peers, or at least among the foremost of them, in terms of care delivery.

To me this would be what would or should happen in any other work sphere. If I had been working on a production line or in a supermarket the management would surely be intimately interested, involved and have stake in seeing a worker who was not performing as they should to understand first *that they were failing* then *how they were failing*, then to ensure over the next few weeks, through monitoring, that they had taken onboard lessons from this failure and to provide them with a clear strategy to improve their performance to bring it up to standard. Ongoing from this further development plans should be put in place so that such a worker could then move on toward further improvement so that that they would not just be performing adequately but, from the experience, to actually perform above average.

This to me would be the only way such an (in)formal warning and admonishment of past bad practice could be addressed and then used as a springboard to turn a failing worker into a better one, indeed a better than average one – a high performer. In my personal case I was seeking guidance – guidance of how improve, feedback on if and how I had improved and then a clear steer to keep improving and becoming a better carer. However much I rejected the complaints made against me what I was desperate for was some resolution that would prevent me from becoming totally demotivated by the experience.

If this were true in a work scenario based around "things" instead of people, how much more important should this be in a caring profession? If it was thought that I was failing and that the result of this failure was a direct negative impact on those I was charged with caring for – that in fact my actions were deemed, to a lesser or greater degree, as harming or degrading the existence of vulnerable individuals - this sort of process would surely not only have been wise in respect of good management practice but that it was *a moral duty on behalf of management* to see that such failures were addressed and turned around. Yet it had been totally and completely lacking, absent, MIA. I had been taken apart in one instant and the next left with no way of knowing if the perceived failures were ongoing or if they were fixed. That dumfounded me.

It shouldn't have done though as this is the way care in the private sector is – CW's can be either bollocked or ignored; sacked or humiliated; disregarded or seen as inconvenient – these are the two poles that mangers and employers dwell in, all other latitudes hold no interest for them – CW's are not seen as integral to the care process, as vital exponents of care delivery, but as units of labour through which service users are cared for and so keep paying out – if service users are biological ATMs CW's are conduits – the wires - through which that money keeps flowing – that and nothing else. There is no interest in the private care sector in seeing CW's develop as individuals or as carers unless there is an intimate and clear financial interest in it. There is no dedication - in fact no will - to improve individual performance or of recognising good carers and good care, there is only the attitude that they must be kept in line in order for them to "transmit" the transaction of money from service user to service provider, whether these "transmitters are copper or fibre optic makes no difference as long a signals keep coming through.

From a personal point of view what I wanted from the whole dreadful experience was some feedback so I knew that I was not walking on eggshells the whole time, crippled in working with others, terrified that any actions in the process of caring that I undertook would subsequently be used against me in some way, and to regain some self belief that I was not the person - I was not the carer - I had been made out to be. Selfishly this was not for the benefit of anyone else, what I wanted was an affirmation of my work for my own sense of worth, of self respect, of sense of self. I wanted some inkling of the knowledge that I indeed *was* a caring individual, that I was a good carer and that the care I was delivering was enhancing rather than detracting from the lives of those I cared for. In short I wanted to begin to believe in myself again.

What I *needed* more than simply *wanted* though was a recognition that my work was satisfactory so that should any further complaints arise I could point to the fact that I was working in the same manner as that in which my work had been monitored and found without blame or failure.

From a professional point of view I was left wondering one of two things – first, as described above - if the complaints made against had been taken so seriously, why was there no follow up, why was there no apparent strategy for tackling my perceived failure, why was I one moment given such a shafting that my backside resembled a Japanese flag then next totally ignored? If this was standard practice, I thought, no wonder so many of those around me appeared so demotivated that they could hardly raise themselves *to care at all.* If this was the way that the whole care sector was run I wondered how anyone could possibly survive and still keep the passion for care so necessary to decent care delivery.

The second thing my mind switched to though was infinitely more troubling for me in this particular instance – if these complaints had at one moment such vital import and yet six weeks later were treated as if nothing had really happened, had something occurred in the meantime to blunt the force of these complaints?

I put this second thought down to paranoia – that was until people started to talk to me again.

It took a while, my period of effective social exclusion lasted for about that whole 6 weeks with only a little thawing in the widespread chilly attitude. Then it suddenly collapsed altogether.

The reason?

The deputy manager had been temporarily "removed" to another home due to claims of "inappropriate" relations with young female staff – this gossip, it transpired, had displaced me as the topic of "water cooler" interest, it was the talk of the home. Having just had my own bruising experience of trouble and then the attendant gossip I cut short anyone who tried to talk to me about the deputy manager's situation in the most polite way possible - I didn't want to be a part of something – idle gossip – or more precisely, the gossiping of idlers - as it was nearly always the carers who did damn-all who were the most prolific gossips - that had so torn me apart, and, I guessed, that if there was indeed any credence to this gossip then only a grain of truth was preserved in it.

Besides it had nothing to do with me apart from the fact that I was no longer the focus of increasingly fevered and wild speculation – in fact I derived some guilty relief at the whole process because I was now sliding down the league table of idle assassination by hyperbolic repetition of fiction, to be replaced by an infinitely more prurient one.

I remained separate from it until one carer who I still trusted took me to one side and asked if I had heard about the rumours.

This carer had been the one bright corner of the otherwise unmitigated dark tunnel from which I was only just emerging; they had not long started and I had worked with them for a short while (my banishment of working alone concluding as other team members left and when, quite suddenly, my "experience" in care had become stunningly applicable to working others again as I was "expected" to show this young - totally inexperienced in care - CW the ropes – the fact that I, not long ago censured for my working methods, was asked to do such a responsible task again begs the question – soon to be answered – as to just what the whole recent episode was really about) – and they were a dream to be paired with. Enthusiastic without being over-eager, driven without being headstrong, and, what was most important to me, selfishly so, unwilling to listen to idle talk, leaving judgement up to their own experience. They were to prove, just as I was down on my knees, so incredibly supportive that I can never be more grateful to them - I firmly believe that in this time they saved me from a full-on nervous breakdown.

Over a very short period I built up a decent working relationship with them and knew them not just to be mature beyond their years – they were still in their teens – but also, considering they had never worked in care before, a natural carer; instinctively knowing what carers far more mature in years and far more experienced in time in care, took a working lifetime still to remain ignorant of. They had boundless energy and a willingness to learn that re-energised my own caring.

As they began to relate to me the details about the "scandal" regarding the Deputy Manager I tried to say – in the most delicate yet forceful way I could – that I wasn't interested in gossip. However in response they told me equally forcefully that this wasn't gossip – they were facts - as they had direct experience of the matter first hand and also had direct knowledge of the nature and identity of the individuals who had made complaints about the "inappropriate" conduct of the deputy manager – further, and this is what made me listen instead of walking away, they told me – and I recall the exact words -

"You need to listen as this involves you too."

At this I was stopped in my tracks. Some of the things I knew, most of it though I didn't.

Over the last 6 to 8 months since the deputy manager had been working in the home he had set about doing two things – marginalising, and in some cases removing, CA's he "did not like" and replacing them with very young – between 17 and 20 – and very attractive female carers. His chosen method of removing CW's he "did not like" was to lean on these young carers to provide statements about such CW's in order to fabricate cases against them - his chosen targets were all men. I happened to be the last man standing – apart from the Deputy Manager I was the only male now working in the building.

This CW - who was young and attractive but who had not been recruited by the deputy manager - had, not long after starting working with me, been asked *directly by the deputy manager* to provide a statement against me alleging that I had abused resident. This he had asked for as recently as in the last 2 weeks; in return he had made some vague promise of further favourable treatment.

At this sentence I was literally rocked back on my heels, but nothing prepared me for what came next.

They went on to say that they knew that *all* the other statements made against had been at the solicitation of the Deputy Manager. I asked *how* they knew this and they replied that some of the people who had made the statements against me had told her themselves that they had done so at the deputy managers insistence; more than this though, these individuals had been told by others that they too had been "asked" for statements against me.

Critically two of the statements that had most hurt me personally had been purloined in the most ruthless of fashions. The go-along-to get-along carer had been coaxed into making a statement by a sustained period of gentle pressure that included constant text messaging - a common theme it transpired – they had also been told that making a statement against me would be beneficial to their career. In the case of the carer who I had worked with for some time they had been promised that on the back of a statement against me they would get an expedited transfer to a home closer to where they lived that they had been asking for for some time.

Suddenly everything started to fall into place, I knew that the other 2 statements made against me wouldn't have taken that much effort to solicit – but even the details of these were revelatory. The fantastically lazy CW - who had been the source of one of the other statements against me – had made the fact an open secret – open to everyone else, secret still to me alone it seemed – and this was because she had bragged to other carers how she had managed to "dump me in the shit" and that she had been "glad" to do so.

The gossiping trouble monkey too had let slip – how could they possibly keep it secret? – that they had got me into hot water to get themselves out of trouble. In this case they had run over and seriously damaged a resident's foot with a hoist while attempting to hoist them alone – an action that demanded 2 carers – the deputy manager had promised to make this issue "go away" if they provided a statement against me. He got the statement, I got the shit and the written record declared that this resident had damaged their foot having got it caught in the bedrails – no hoist injury – it didn't happen.

Why?

Because the written record said it didn't.

The carer in question here just couldn't keep their mouth shut – any sensible individual - knowing that they had colluded in creating a false reason for an injury to a resident that they themselves had caused through negligence - would have kept their mouth shut, but, when the Deputy Manager found themselves on the wrong end of a number of statements themselves the gossip had taken it upon herself to relate this tale to prove that he was an "alright bloke."

What had eventually run the Deputy Manger into the sand himself was repeated text messaging, unsolicited, of a number of young attractive female carers. Every carer had to submit a mobile number as well as a landline so they could be contacted to cover for sickness, this was kept in a nursing file but in a common area. What the deputy manager had been doing was looking up the mobile numbers and using them to text messages of an increasingly personal and intimate nature. This had been going on for some time and had been the cause of the go-along-to-get along carers' departure. Apparently after they had made the statement against me the Deputy Manager had stepped up the texts and also his behaviour at work. More than one carer said they had witnessed the Deputy Manager being "overly familiar." With this CW. Overly familiarity is what brought the whole matter to a head though when he – noticing a stain on the uniform of one of the young carers over the breast area - had asked if they wanted him to lick it off.

Oh dear.

Unfortunately for the deputy manager this carer had kept many of the texts he had sent and after the "stain" incident showed them to their mother who was a senior nurse at a large local hospital; she duly came in steaming mad and the Deputy Manager had that same day been sent "on-secondment" to another home.

After all this had come out from the carer I was working with I was open-mouthed. She too, she told me, had been subject to unwanted text messages but, on refusing to provide a statement about me, they had stopped.

I didn't know what to say. First of all I just could not believe the scale of it, the organisation, the planning, the plotting, and then descending on me and my reeling mind like a brake shoe in an emergency stop – Why?

Why?

Why?

Why?

Why? Why? Why?

Over the next few weeks I racked my brains as to why the Deputy Manager would want to see me canned, what had caused it? I thought over every action I could remember that could even tangentially apply to the Deputy Manager. I thought of every interaction with him that could possibly account for his animus and I could still think of nothing, nothing that is except one thing. The "glans episode". This was the only thing I could possibly think of. Had that so affected this Deputy Manager, had he harboured a grudge, a vindictive urge to see me purged, because of a conviction that they saw me as some sort of a threat? Or was it simply a case of wanting to see someone who they had thought had professionally embarrassed them put to the sword.

I couldn't think of anything else. And then I remembered that flash/stare.

I thought I knew how shafted I was at the time, but really I didn't have a clue.

Until now.

Had the *real reason* behind all the complaints come to light with the "inappropriate" texts and behaviour? Was this the reason that there had been no follow up review of my performance after the carpeting? Was this why the severe reprimand had been followed by wilful indifference to the whole matter? Was it that instead of offering an apology or acknowledging that the case against me had all been a "put-up job" it was easier instead just to brush the whole lot under the carpet along with me?

It surely couldn't all be coincidence, couldn't all be totally unrelated.

Initially, after hearing these words, a wave of relief swept over me. I felt as if I had been vindicated and the horrible burden of the deep and uneasy knowledge that perhaps I had been an insensitive and almost brutal CW was lifted of my hunched shoulders. For the first time since the incident I felt like I had done nothing wrong, that the long chain of guilty self-reproach that had clattered around my fettered conscience had been cut loose.

But then, over the next few hours, another sensation replaced elation, one of being still horribly trapped by facts. The truth was that as long as the statements stayed on record I was still vulnerable – the made, however purloined, still remained on my personal file - so how would these statements play out in some adverse turn of events in the future?

Although I felt I was liberated it was only a private liberty. I could prove nothing as to the "fitted up" nature of the complaints made against me, and those who knew the truth were hardly likely to march forward now, either way most of them had moved on or out anyway so had probably not given me a backward glance. In all the most meaningful ways – apart from the *most* meaningful *which was to myself* – I was in a worse place than I thought I was in before. The Deputy Manager had been moved for the whole "inappropriate conduct" allegations to blow over, however he would, in the end, be back and no doubt looking for scalps to cover up his own indiscretions now.

I started looking for another job, it seemed the only sensible thing to do given the circumstance.

Yet had been mentally shattered by the whole turn of recent events, and the future looked as uncertain as the recent past had been harrowing. Did I really want to work in a line of work where denunciation and fabrication were the norm? Where personal reputations could be shattered upon an instant with little supporting evidence? Did I really want to work in an environment where a seething mass of personal resentments were justification for personal vilification?

It may be thought that I was being overly pessimistic that *all care organisations are the same*, yet the experience of others will prove this is far from the case. Further it shows how good care is actively mitigated against and those who still provide it operate on a knife edge of reputational risk and ultimately career death.

In This Job You're Innocent Until Investigated

The preceding chapter throws up a number of issues that I believe need expansion of in order to understand the febrile nature of the care work environment for CW's - from the broken reporting structure of the care needs, additional requirements and presentation of service users from CW on up to clinical professionals, and the lack of interest, indeed antipathy, of those professionals as to what – often important - information CW's tell them; to the distorted "through the looking glass" nature of complaints procedures in care – where what is good is wrong and what is wrong is good - against CW's and the ever present potential for such complaints to "come out the woods" for the most perverse reasons; to the Kafkaesque way in which those complaints are handled, judged and sanctions applied and finally through to the almost totally exposed situation of CW's to challenge, rebut or answer such accusations.

We shall come to complaints shortly, but first, as it is the most serious of disconnects in care, is the often difficult relationship that carers have to clinical professionals.

CW's are used to being regarded a little else than neo-colonial coolies in the "jungle" of care work, of being assumed to be - by a majority of relatives - the dregs of society with little compassion, understanding or intelligence; but it would be expected that their standing would be somewhat above this basement level "inside" the care environment; it would be expected that the clinical professionals they most often work with – nurses - who see and can judge their competence day after day – would have different – more positive - perspective. This sadly is not the case. Few are the nurses I have met and worked with who regarded CW's as anything else but an irritation or an inconvenience; more than this though they often regarded any proactive carers - ones who fully utilised their skills and sought to extend their knowledge beyond the bare minimum - with suspicion and often outright hostility.

Many of the manifestations of this fraught relationship have been thrown up in the preceding chapters in one form or another, but here I would like to note how critically dysfunctional this makes what should be a streamlined and reactive "chain of command".

One of the best nurses I worked with called CW's their "eyes and ears", critically important to their awareness of a change in the condition of any service users and utterly reliant upon them to report back not just what they have observed or noted, but also to employ - borne by their more extensive contact with service users - their intuition from these observations as to what might be any underlying issues regarding any changes.

For example this nurse was told by one CW that a service user was more confused than usual, was apparently having visual hallucinations and was accepting very little food. The CW also reported that these were the exact symptoms they had noticed before in this service user prior to them being diagnosed with a Urinary Tract Infection (UTI). This nurse valued this information, obtained a urine sample, dipstick tested it and found this to be more than likely the case, and on the evidence of this they requested a doctor's visit and recommended antibiotics. In brief this nurse not only was receptive to the raw information that was passed on to them but they also paid attention of the interpretation of the symptoms noticed by the CW and acted swiftly upon them. This service user was on a course of antibiotics for a UTI within 24 hours.

In contrast – and by way of another example - on one night shift a very frail, very elderly service user, with a history of persistent chest infections, presented with a very "rattley" chest, they too were more confused than usual and their sleep pattern was disturbed. The CW caring for this service user informed the night nurse who came up to the floor they were working, looked in on the service user for all of 30 seconds and declared they could see no difference from their usual presentation. The same CW noticed the same "rattle" the next night and reported it, nothing doing; it was only when another nurse was in on the 3rd nightshift that the information was acted upon – this nurse went up to the floor and observed the service user for 10 – 20 minutes and requested of the dayshift the next morning to call out a doctor – a chest infection was duly diagnosed and antibiotics begun as a matter of urgency – 48 hours later than they could have been – a window that in such a very elderly and frail service user that could have meant – and I don't exaggerate here - the difference between life and death.

Examples similar to the second case were so persistent and occurred so often that frequently CW's simply stopped passing on their observations to nurses and thus failed the service users they were supposed to be caring for.

The above example – and the trend it is indicative of - may be considered as evidence of George Bernard Shaw's maxim of all professions being a conspiracy against the laity - that nurses in care see themselves as possessors of a prized and arcane knowledge beyond the ken of all but themselves; or it may just be that they view CW's as dimly as the rest of society. What is beyond doubt though is that the routine disregard that nurses have for CW's seriously impedes good quality care for service users. Further it can actually be openly damaging to their wellbeing and welfare, and, in extreme cases, can lead to avoidable deaths. The early detection and action upon any changes in a service users' presentation is vital when dealing with individuals in an already fragile state of health and the store of knowledge built up by CW's can be equally vital in diagnosing an issue swiftly and efficiently.

Shaw's above mentioned maxim may be the reason why proactive carers who are willing to push the boundaries of their knowledge and broaden their aptitude are treated with such hostility. There seems an endemic suspicion that those CW's who appear, or actively try, to advance suggestions and interpretations of how and why service users' presentation and actions may have changed are somehow "getting ahead of themselves" or have "ideas above their station" or it may just be that both health professionals and managers fear – God forbid – the development of independent thought in what they see as little more than care automatons.

Whatever is the reason it is routine that such proactive CW's are almost always slapped down – preferably in the most public way possible – and often deliberately ridiculed and humiliated. This leads to total de-motivation in the CW workforce and worse – having been serially ignored/effectively punished for trying to do a more than simply adequate job good CW's just stop making vital observations or of passing them on; this is what happened to me.

Bruised by my encounter with the senior nurse/deputy manager I lapsed back into only passing on the most banal of information, of simply relating what I observed; and even this little input I found was also similarly ignored. As a venture in purposelessness, the objective assessment and recording of the presentation of service users by CW's without the slightest notice being taken of them makes what is already a difficult job into something similar to Dostoyevsky's observation of the ultimate spirit breaking punishment - the task of having someone dig a hole then be asked to fill it in again only to be ordered to start digging it all over again.

Camilla Cavendish in her report into the role and purpose of CW's thought she had a solution to this problem by suggesting a clear and defined career progression path that would allow go-ahead CW's to advance in both expertise, knowledge and position. This was a decent point but arguably missed the target. Most CW's like being CW's simply because they like the job of caring, they like the contact time they have with service users are fulfilled by the knowledge that they are doing a vital job that enhances the lives of those who otherwise would see their existence horribly truncated in value and quality; many do not want either to become nurses or to have extra responsibilities heaped upon them, they do not want to be pushed further up some career ladder or to seek advancement other than extending the skillsets that are germane to their role as carers. *What they do want* however is for their position *as* CW's *to be valued in and for itself*, they would like to be acknowledged as being "on point" in assessing the wellbeing of those they care for and their observations to be taken seriously and acted upon; in short what they crave, to an individual, *is to be recognised.* This goes back to what I have had cause to repeat again and again – with reason – that the status and standing of CW's is currently not - but should be as a matter of urgency - enhanced; instead of being seen as mere adjuncts to care, mere conduits of care delivery, they should be seen as the most important asset in care and accorded the respect and recognition that this denotes.

Once more I issue a *cri de coeur* that CW's should be professionalised and given a chance to similar to that of nurses. This often repeated mantra would also go some way in ameliorating the way complaints against CW's are made, treated, and – ultimately - handled. At present CW's – simply slaves in the care jungle – are liable to be whipped and scourged into indolent indifference by their vain "owners" simply for doing their job.

The fact is, as things currently stand, that as a CW you are only as "clean" as "good" and as "blameless" *as other people say you are.* There are no hard and fast rules you can grab on to – you cannot reach for standards by which you can be objectively measured, instead everything is subjective and an entirely innocent or well-meaning action can be spun out of all context and into a disciplinary just by its selective relation by others. Against this onslaught of accusation by slander there is very little an embattled CW can do.

No one who enters care is ever quite fully aware of how perilous a position it is that they occupy and how the forbidding environment is in which they operate; at any moment a word out of turn here or there to the wrong person, an action that is in the least part open to interpretation, even the impressions perhaps inadvertently made on others – service users, their families, other CW's - can result in you being neck deep in trouble and sinking fast - as the previous chapter has hopefully shown.

In some ways it could be argued that all health professionals find themselves in an equally as arbitrary environment in which the topography can swiftly change – one moment at a mountain top, the next dead in a ditch; the main difference is – and this is a point I have returned to again and again - is that nurses and doctors have the backing and support of a professional body, that while it has its own rules and standards by which they are measured that can often times seem just as confusing as those operating in the "jungle" of the care environment, at least offer a modicum of protection. A manufactured complaint is much more difficult to uphold against a nurse in care because they are able to appeal to a professional body. CW's do not have one so cannot.

Another small example will illustrate this.

A nurse I worked with, and whom I had more than a couple of run-ins with due to her rather offensive nature, at times bullying attitude and her more than just latent arrogance, but whom I nevertheless respected for their professionalism and experience, found themselves, with no warning or intimation, suddenly suspended.

Like me the whole nature of the complaint made against them were based on statements made by a couple of care professionals with an agenda. An outside (but internal to the company) individual was brought in to investigate the complaints and on examining the evidence declared there was no case to answer, however the home manager still insisted that this nurse be given a written warning as to their conduct – *even after they had cleared of any wrongdoing* - at this the nurse brought in a Royal College of Nursing representative who stated baldly that if the manager did want to issue a written warning then it would be open to challenge - supported by them- at which point the whole issue was dropped like a hot brick.

While this shows that nurses can be open to capricious and malicious complaints it also shows that ultimately they have recourse to their professional body for expert, professional and - as a cast iron backstop - legal representation.

Care, as I found to my cost, can be an intrinsically vindictive place to work. As I noted early on in this book it is sometimes hard to distinguish carers from the service users in terms of the mental issues displayed by any number of them – a situation, as I have described, I am not placing myself above. This translates into the factors that make care a particularly toxic place to work; the merest slights can result in the most acrimonious of results because of the, frankly, "complex" nature of the majority of the workforce.

As I have shown above, merely asking someone to do the job that they are paid for (and, more worryingly, to ask them to discharge not just their duty of care as defined by workplace practices but the *human moral obligation* incumbent upon caring for the weak and vulnerable) can have consequences – the lazy individual of a CW who I asked to actually do some damn work apparently saw this as a mortal stain on their escutcheon that demanded payback.

And care produces the most egocentric, not to say sociopathic, class of action - nothing stands in the way of radical self interest – if there is a personal pay-off of in making an inflated or just plain false allegation that can advance someone's own interests then, in the main, they have no compunction in carrying out the most morally dubious of actions.

This though is not the most troubling aspect of care and CW's. As I have already related previously I have encountered episodes – too numerous to simply put down to coincidence or unknown factors – whereby malicious complaints have been made *for no apparent reason other than a will to do harm.* This for me has to be the most difficult aspect of care and working with other CW's that I experienced, simply because it is beyond rationalisation.

While difficult to take, a malicious complaint that offers *some advantage – material professional or personal –* to another CW is easier to fathom because you can see *some angle in it for them.* Those made out of apparently sheer vindictiveness, with the sole aim to see others suffer are difficult to understand or quantify. What is infinitely more worrying - and as again I have already had cause to mention – is the wanton cruelty that this betrays which speaks of a mindset of viciousness that sets alarm bells ringing as to the suitability of such individuals to be in contact with vulnerable service users. If the very people relied upon to treat often difficult service users with respect, compassion and dignity can display and manifest such a nefarious streak would you really want someone with such a flagitious character aspect looking after your relatives? No way in my understanding of a caring and compassionate attitude to others that has to be the most fundamental requirement for CW's.

The main problem in care is allegations can, and are most frequently, made against the most efficient and effective carers that often result in their removal from a workplace or from the care sector as a whole. This is so simply because of a matter of statistics - the more proactive a CW – that is the more work they do – the more this exposes them – exponentially - to the potential for dubious claims of wrongdoing and the seriousness of those claims – put simply by way of example – an effective carer who attends dutifully to, say, service users personal care, will be exposed to more risk of potentially very damaging accusations than someone who simply lets service users sit in their own faeces and urine.

This has a multiplier effect in the degradation of the quality of care available for service users as not only are they denied the services of effective CW's - as effective CW's are most likely to find themselves on the wrong end of a disciplinary or the bullet - and are therefore driven out of care in one way or another - but it also means that poor, lazy and/or incompetent carers remain in care while those good and effective CA's are driven out.

Finally there is the issue of how complaints are handled in care – the Orwellian, or perhaps more aptly Kafkaesque – nature of most complaints procedures. When I was dragged into the manager's office - as described in the previous chapter – I was neither told the specifics of the complaints made against me or their source, therefore I could mount no defence against the accusations because I simply did not know the substance of them. Not knowing the source of them either I could not challenge any interpretation that could possibly have been put on my actions.

While the confidentiality of reporting may be understandable – workers, especially in care, need to free from any possible bullying or intimidation to withdraw complaints and be shielded from possible consequences after - this safeguard is however frequently not employed as intended but instead used as an impenetrable veil by which malicious accusations can be made free in the knowledge that that the identity and source of the complaints can remain hidden. This abuse of process is equivalent to kicking a blind man in the backside.... at night. I remain certain that if I had known the source of at least one or two of the accusations *at the time I was accused* I could have offered some background to it, or even have rebutted the claims as totally bogus. However on this point I remain ambivalent – while it does offer protections for the vindictive and malicious I believe that confidentiality of complaints is essential to the freedom of all those working in care to draw serious abuses to managerial notice without fear of retribution or intimidation.

However on the score of masking the very circumstances and nature of the complaints I have no such divided feelings. It is one thing to protect the confidentiality of sources by being non-specific or overly descriptive of actual incidents in question but quite another to have those same incidents made so unnecessarily opaque that it denies the subject of the complaint any grounds on which to mount a defence, rebuttal or add context – to be censured for an event that the subject is essentially totally in the dark over defies any sense of justice or equity.

If my brush with such complaints was difficult though it is nothing as compared with two other incidents that go even further in showing Orwellian character of care.

One CW I had a close working relationship with that endured for over 2 years – and who ironically went on the to be source of one of the complaints about me – found a hand delivered envelope delivered through their door one evening as they were getting ready for a night shift. It tersely informed them that a serious complaint had been received about them, they were forthwith suspended until further notice while an investigation was conducted, that they were excluded from entering their workplace, or any others in the same company, that they were – under threat of immediate dismissal – banned from contacting the place of work or any employees of the company and that in due course they would be "invited" to attend a disciplinary hearing at a date yet to be set.

They were neither informed of the nature of the complaint, its direction or any intimation of any of the circumstances surrounding it, further, as they were banned from contacting the workplace, they could not ask about independent representation or the possibility of making a defence against the complaint prior to the disciplinary hearing.

At work we were told that the CW was "on the sick" although, as I was to quickly find out, this fabrication was somewhat built on straw as – and in a foreshadowing of my experience – it turned out that the substance, facts and details of the complaint were common knowledge throughout the care home. In this instance the "leaking" of information was fortuitous for the suspended CW as the complaint turned around an event that was said to have taken place on a particular date. Because of the leaks I knew that the complaint was groundless as I had been working with that CW at the date and time of the supposed infraction. However I was now in an impossible situation, as I was not supposed to know the detail of this complaint I could not submit the information that could exculpate them. This was a ridiculous state of affairs but is typical of complaints in care – even when other employees have information that could clear a colleague of malpractice they cannot revel it without getting themselves and possibly others into a disciplinary situation.

This would be farcical if it didn't kill careers and reputations.

For days I agonised over what to do about the situation and in the end stuck my neck out and contacted the suspended CW and told them that in the hearing they could call on me to offer a defence as I knew I had been with them when the supposed offence took place. This put me and my career at risk but I deemed it worthwhile as I did not wish to see this – very good and competent – CW dismissed unjustly for something they did not do. The suspended CW did call on me at their hearing and I could offer the exculpation and all was cleared up in an afternoon.

It could have been very different though, in fact rather like what happened to a friend of mine.

They went to work as a senior carer/deputy service manager in a supported living environment for minors. When they started in their job they found the typical "care worker led" culture where CW's could and did do pretty much as they wanted; their work attitude and aptitude was a mixture of laziness, incompetence, dereliction and neglectful disregard for those they were meant to be supporting, combined with the usually associated heroic levels of sickness.

A weak manager and poor oversight had led to this state of affairs and, by any measure, the service was falling short or even failing. Indeed the service was only barely passing muster – and inspection – the last inspection having ordered several remedial actions to be put in place under threat of serious sanction. That it was not totally failing was due to a very small number – and by small I mean one or two – CW's who were constantly picking up the pieces for those too inept or uninterested in actually caring.

On commencement of their appointment my friend found out that their actual employment had been commissioned by the service provider for the very task of pulling the service into line, so they set about rectifying these failings. They were "on point" in ensuring that, for a change, CW's did the job they were paid to do, in this they experienced severe resistance and so had several causes to administer stern warnings/admonitions/reprimands for poor work standards that included leaving service users unsupervised, leaving them under – or totally un – stimulated, failing to adhere to proper policies and procedures in respect of supporting service users, failing to follow care planning and persistent ill-explained or unjustified absenteeism.

Knowing my friend as I did they found this process very difficult; not combative by nature what motivated them into unavoidable confrontations was the fact that service users were being badly failed by the care team and this was having a direct impact on their quality of life (it was not uncommon for service users to go for days without even being taken outside the facility, even though social activities were central to the care planning of nearly all of them) indeed what my friend was seeking was to create a more professional and cohesive care team whereby care delivery would be consistent and integrated and to pull the failing CW's up to the standard of the one or two good CW's.

Suddenly, 3 months after their appointment, they found themselves one morning - and out of the blue - barred from work and prevented from entering the premises; they were told – and this may have a familiar ring by now – that they were being investigated for several serious complaints - in fact so serious that even the supervising social work authority (as this example took place in Scotland there is a different oversight and clearance system to rest of the UK – the supervising body being the Scottish Social Services Council –SSSC – which also issues clearance and registration to work in the social care field) had been brought in to look into them, as the allegations were judged to potentially consist of not just internal (to the service provider) misconduct but also professional misconduct. Of course they were not told what these complaints consisted of, when they were alleged to have taken place or who they involved, only that they were excluded from the workplace, forbidden to contact anyone connected to their place of work and would be called upon in due course to "answer" the "charges". For good measure their registration to work in the social care environment issued by the SSSC (the equivalent of the DBS in the rest of the UK) had been pulled indefinitely until the mater had been fully investigated.

Finally they were "called in" – the charges, such as they were – were either ridiculous – one for sexual harassment by someone they hardly ever had any contact with – bullying, intimidation, and mismanagement. It was clear from all of these that a "cabal" of the uninterested disaffected had banded together to bury try friend in complaints which, piled together, provided a formidable wall of retribution. At their hearing my friend was not told of the circumstances or the times and places when the actions under question took place so it was impossible for them to rebut them; the short hearing abruptly terminated with the blunt choice of resignation or sacking. Having no representation or even the basis on which to answer the charges and to avoid a sacking which they thought would place their registration at risk my friend chose to resign.

Unfortunately, having resigned before the charges could be fully judged, one the of the "cabal" took it upon themselves to contact the SSSC to say that the charges against my friend had not been fully resolved as they had resigned before they could be properly investigated. The SSSC then viewed the case as having not been concluded so continued to withhold my friends' registration - meaning they could not apply for *any* job in *any care field or one associated with needing SSSC registration.* In brief they were screwed.

Having resigned they could not apply for unemployment benefit or any other associated state benefits, they were in despair. Eventually they applied for a job overseas and left the country – a direct result of the impossible complaints process. Let me run that past you again – *they had to leave the country directly because of the manufactured complaints against them.*

If any definitive case were needed as evidence of how totally messed up the complaints procedure is in care work then this is surely it.

While all these examples show just how disastrous the complaints procedures are in care they also point to a larger paradox and contradiction that runs right through care and which threatens the very basis on which it is delivered.

We, as CW's, are drilled constantly of the need to treat all service users with dignity, respect, individuality and the freedom to make their own informed choices regarding their care and lives (many outside of care will ask why this has to be drummed into CW's as dignity, respect and liberty seem the basic requirements on which we should treat others in life in general) yet in our places of work we as CW's are constantly treated by our employers with disrespect, subjected to all manner of indignities, and have our freedoms as individuals increasingly curbed; we are treated little better than cattle – to be poked and prodded, pushed and crushed, corralled and penned in, and then oftentimes hung out and bled dry.

The paradox and contradiction then is clear – how can a workplace in care, or indeed any workplace, foster an environment *of care* when staff are treated in such a careless and vindictive fashion. The truth is it cannot; dignity, respect and choice begins with the staff – if they feel empowered and respected they will pass this through to their care, without it the whole edifice crumbles.

And this in a nutshell is where my thinking in care and about care ended up. I realised I was at a turning point, I had a career - not just a job - decision to make, and more was at stake than just how I earned a living - there was also the way the job had been slowly eating away at me from the inside – unnoticed until this pivotal stock taking – corroding me on a more fundamental moral level.

This wasn't just about the environment that I worked in, not about complaints or statements or carpetings by distant managers, this was about the fundamental basis on which the care system I saw around me functioned.

At every level I saw the *care-less* nature of the of the social care sector, from the highest points of government and policy, through the oversight of bodies charged with policing the sector, descending through various care providers and their managers and on to the the very care workers who actually delivered care; at every level it appeared to me - with very few exceptions; honourable, hardworking, dedicated and truly caring exceptions – that care was absent from almost every calculation, either people, institutions or businesses were indifferent to the care we were supposed to be providing or else they were driven by quite different motivations – for money, for advancement, for personal gain or simply for an easy life in work; nothing made sense to me. I looked around and saw a system that was failing and that I was part of that system.

I felt like I was slowly rotting from the core outward, worse I saw for myself the future that I saw in others around me, one where I became disengaged from caring, where in fact *I stopped caring.*

I looked ahead and didn't want that future for me.

So I left care. I became one of the many displaced refugees from it. All my ideas, perhaps all my dreams of "making a difference" seemed now so remote and naïve as to actually grate on me, like some childlike ambition of being an astronaut, they seemed simplistic, silly and stupid.

Yet the questions remained, I could not lose them. And I could not keep them quiet. Someway, somehow I had to get them out, I felt the urge to tell people exactly what social care was like from an insiders perspective, and not from a managers perspective but from a care workers one, the voice that has been missing from all public discourse and debate about social care and especially about care of the elderly, thus the reason I am typing out these lines now - nearing the end of what I hope will wake the some few who read these words up, for them to really understand what care is like, where it is now and what is being done in the name of that widely abused word.

Care.

What do we mean by it?

Social care.

What do mean by it?

Respect, liberty and the chance to live a full life right to the end.

What do we mean by it?

I hope in the preceding I have offered concrete examples, not least my own, of what these terms mean in the actual practice of them rather than just a form of words, platitudes to be trotted out thoughtlessly.

The picture is not pretty but it is true.

I could end here but two vital questions remains with me that I must release from my own mind – where is it that we are going with care, with social care and what can or could be done to make care an actual reality rather than misnomer.

These are the questions I shall come to next.

CHAPTER XXII

In A Handcart

"The abuse scandals that make the headlines.........make us lazy and complacent. They divert us from the low level neglect and insidious undermining of [care home] residents' autonomy that is far more frequent."

Article from The Daily Telegraph 23/2/15.

In the preceding I have covered aspects of care and the way the care system is orientated and organised that reflect the "low level neglect" that the above article speaks about. My experiences, as stated at the outset, do not reveal spectacular and so "newsworthy" incidents of abuse and suffering, what they do show is that subtly, daily and widely care home residents suffer "care" that not only fails to reflect and respond to their individual needs and requirements, let alone their wants desires and wishes, but also leads them into a miserable existence that betrays the life they have formerly led by making the very last months or years of it ones that indignify their status as sentient (now matter how truncated that sentience is) free and personally independent human beings, they become just a physical constellation of care tasks that need to be carried out *upon them*, in short they become a job.

The above quoted article goes on to note –

"The heart of [the] problem is the dehumanising aspect of placing people in institutions and treating them as commodities rather the human beings."

This is the entirely accurate and what I have described at length in the above – the commodification of care - has led ineluctably to the commodification of individuals – while CW's may regard residents as a barely human "job" care home operators regard them as cash cows, their job, as they see it in reality as opposed to in the buff of the glossy brochures that they produce to lure the unwitting into their net, is to provide the minimum of care necessary in order that they may be retained in their racket (yes, racket, let's call it what it is) and keep paying out the required lucre. This why they employ carers on minimum wage, why they impose regimes that reflect the needs of the business rather than the needs of their clients and why they race to throw up ever more salubrious establishments that dazzle the eye in order to distract from actual things that really matter – such as care that displays any shred of humanity and social obligation.

This is why care has become a business.

Who would have thought it back in the post-war years when the welfare state was established, who would have considered this the bright future of the 21st century when the architects of universal social care coverage set out to liberate the masses from the tyranny of the market that decided what level of care you received based solely on the financial means you had at your disposal? Did they really envisage that 60 years later the market would dictate whether you lived in an environment that properly reflected individuality and liberty or one that reduced you to an anonymous "package" of funding (yes they *actually call care funding a package* – irreversibly associating and intimately linking the financial to the individual and so reducing the latter to nothing more than figures on a cash flow) that made you hostage to the pounds and pence you come to represent. Would they be happy with what we have done with their legacy?

Much of the coverage of the crisis or indeed the affordability of the socialised system of healthcare and the attendant public neurosis about the future of it has centred around NHS, acute or local – that is hospitals and GP's – or what some have described as the closest thing the British have to a national religion such is the esteem and love in which the NHS is regarded. Yet the NHS was only part of what the architects of the universal healthcare, free at the point of delivery, had in mind; what they were in fact building was a "cradle to grave" system that would look after *all* social needs, not just ones associated with doctors and nurses, operations and consultations, what they were building was a network of social care – an idea that the state would provide care free from market interference.

What they did build – with admittedly grave shortcomings – was the very idea that what would happen to you in old age would not be a worry as to financial means, would not involve hard choices as to assets and income, would not be a source of gain for a few at the expense of the many and that an ageing generation would not be left with a potential lottery as to what type and quality of care they would receive at the point of their most maximal dependency and vulnerability.

What took 20 to 30 years to build – the idea of state provided social care - has, in a equal period- been as throughly dismantled as it had been constructed. Since the 1980's we have unloaded the care for our elderly, mentally ill and those with learning difficulties onto a private sector that has so sedulously robbed people not just of their money but of the quality of their lives; what we have bequeathed to posterity is the worst of the old system – the institutionalised nature of care – whether in the community or in actual buildings – that denies people their autonomy whilst basing this care on whatever money they have available – whether through state aid or their own financial assets. What we have today is an empty concept that the state will always come to the aid of those that need it most, in reality we have a system in which, yes, the state will aid you *but*.......

In short we have care with riders and provisos – that state will aid you.......up *to a point*, but that point is the thin edge of a wedge, it becomes thicker as pressure mounts on financial imperatives; the *but* becomes ever bigger.

And it is getting very big indeed.

In the care sector that supports those with learning or physical dependencies we are currently undergoing a harrowing so thorough that it ranks up there with Christ's tearing through hell in Christian theology. Hoards of social workers are, as I write this, being dispersed through every care service that caters for the mentally or physically disadvantaged and asking detailed questions about their care needs – and questioning carers closely about just what those they care for require, right down to how much they "need" social activities – how many hours? Who with? Where? Can they be supported by one carer instead of two? What benefit do they get from it? Is it necessary?

This is being done not to assess care needs, although this is the title under which it is being carried out, but instead with the sole (unstated) objective of cutting costs. All over social care coverage is being stripped back to the very basics – to see that individuals in care are fed, watered, bathed and changed – *and that's all.*

Support for socialisation and activities are being yanked out from underneath service users and what is being returned to them is simply the process of getting them out of bed and then parking them in front of a TV until it is either mealtimes, times for changing or bathing or time for bed when an extra carer or two *may* be available.

Instead of 24 hour personal support, service users - many with quite extreme or extensive needs - are lumped together with others – often, as I have commented on, with no regard as to their compatibility - simply so that less carers are needed; all this without a thought for what may best suit the service user. In the place of tailored 24hr care are allotted time slots where additional carers are on hand to help with more intensive tasks - and if this doesn't fit in with quite what an individual would like to or not like to do at that time – too bad. Autonomy and Liberty are being squeezed out of people's lives slowly and bureaucratically, and, most of all, clandestinely – *for this process is going on right now,* and yet most people outside care know nothing about it.

And this we call progress.

The situation with care for the elderly is even harsher. Straightened finances are keeping more and more of the elderly with more and more extensive needs - that in the past would have necessitated a move into a residential setting with 24 hour care – in homecare provided by domiciliary CW's.

In all I have noted about the crushing failures in care homes and the shortcomings highlighted in The Telegraph article quoted – the anonymisation, commodification and institutionalisation of individuals – keeping people in their own homes with as much independence as possible may seem like a good thing, and it is; but the pressure on budgets means that the homecare provided is painfully inadequate to meet many of the needs of the vulnerable and disabled, not least the need for socialisation and company that many of those with extreme conditions find themselves doing without, simply because care budgets deem this above the minimum necessary to fulfil their "care" requirements. The Three Plagues of residential care that Atul Gawande wrote so eloquently about in Being Mortal as being abroad in care institutions are now let loose in people's own homes – boredom, loneliness and helplessness are becoming usurping tyrants right within the Englishman's castle.

Domiciliary workers pressed into having only half an hour to give someone their breakfast and medication before being rushed off to the next job is not providing care, it is providing the means for subsistence. Likewise having an hour to give a frail and virtually incapacitated individual a bath is not seeing to their needs, wants and desires any more than the rote care provided in many care homes – it still leaves people at the mercy of harassed CW's and hostage to an external timetable that they have no control over.

And when their needs do become too extreme and have to be met in residential care increasingly the vulnerable, and elderly find themselves in ever bigger and ever more impersonal care homes. This is because cost pressures are making economies of scale increasingly more important, meaning that the move toward larger facilities that cater even less well for individuals are becoming the norm.

Further, the care home sector is ripe for what the market calls consolidation – the absorption of many smaller care companies into a very few very large ones. This is because when times were good and margins fatter a multitude of smaller independent operators of care homes sprang up, eager to sup from the overflowing social cup; they, in the main, could thrive in the provinces, but now that we live in leaner times many of these operators are finding it harder to compete with larger care companies on costs. In the coming years I predict that two things will happen –

First that financial failures of small care companies will increase. This will primarily and detrimentally affect residents as they are left in a virtual limbo – unsure if they will have to undergo the trauma of a move and the possibility that any move may be to another care home less local as well as totally unfamiliar, but it will also critically impact on already stressed local authorities as they will have to step in to provide some continuity of care over at least the short term.

The second consequence is that the "market" will favour the larger care companies; I also predict that in 5 to 10 years' time that only 2 or 3 big care providers - that will have enlarged substantially over that same period due to takeovers of financially crippled or failed smaller operators - will control over 95% of the for-profit care sector.

These care behemoths will not only find it impossible to focus on anything like the individual service user, but that care in such companies will in fact be – if this can be conceived of – more impersonal and, worryingly, more variable; a large company will find it very difficult to impose and maintain an overarching ethos of care – however well designed and intentioned - on individual homes simply because of the unpredictability in the nature of care work – an company ethos may be able to be imposed in the likes of Starbucks but is impossible in the unstable world of the care environment.

What is more a "market" dominated by only a few big beasts will be able to dangerously dictate the cost of care to local authorities and even on a national level - even without a formal cartel or collusion a de facto one will exist; with few choices left open to them and now with no public sector alternative the public purse will have to pay what such big providers demand – we will have ransomed our care with barely hearing discouraging word.

Of course there is an alternative to this scenario of the private care sector milking the national purse and that is that an increasing financial burden will fall on the individual. This is already happening. More and more people in care homes are been "asked" – at pain of losing a place at either their preferred or their most local care home – to pay a "top up" to cover any shortfall between costs met socially either by central or local government or the NHS, and the full cost of the care home.

This I predict will become the not the exception but the rule; further the shortfall between public funding for social care and what the "market" demands will grow ever wider meaning the top up that needs to be paid will become commensurately outsize.

This is not an accident, it is a design. In this fashion we are seeing the total privatisation of social care – as in that social care costs will increasingly, and eventually totally, be borne by the individual.

Insurance companies are already licking their lips at the prospect of selling lifelong policies to meet potential future social care costs – we are moving - in social care at any rate - toward an American system of private insurance for care formerly met by the state.

This may seems alarmist but look at even the most cursory of figures and it appears less outlandish and more inevitable that social care will be raped in this way.

We are facing the retirement and further ageing of the baby boom generation – one that has been raised through the era of fully comprehensive health and social care and therefore are likely to the be the fittest and most enduring of generations that the modern world has seen. Fitter bodies will not be matched by minds so the chances of an explosion in dementia cases – bar an unexpected breakthrough in medical science – is very real and very imminent. If the social care system is under severe strain now it will have no chance against the wall of need that is advancing upon it, at least as it exists in its present form.

The tragedy is this particular manifestation of a "demographic time bomb" has been foreseen for years and yet nothing had been done to counter it. Successive government have opted to stick their heads in the sand rather than plan ahead for what will be perhaps the greatest social revolution in modern times.

In brief where we are going is a bleaker place than where we are now, the problems in social care that we are encountering at present will seem like teething troubles ahead of the challenges that we will shortly face and we are so inadequately prepared that they stand every chance of overturning the very concept of the welfare state.

Yet what can be done?

After all I have written and all the criticisms I have made of social care I feel it incumbent upon me to make some attempt to provide some general answers, so, in conclusion, the next chapter is what this former care worker believes could and would make a substantial difference to social care in the UK.

SOME ANSWERS

Although I want to make this section - that very briefly offers some answers to the pressing questions that have emerged from my story – one that reaches for the stars in aspiration; I also want it to be more than simply an impossible wish list of the vain and incredible; however, as my story is also in many ways the story of social care in the UK as it presently exists – somewhere in the gutter of priorities and practice - then it must necessarily also grasp for the broadest vision in making some conclusions, these may be overarching and grandiose, but they are needfully so, for, without such elements, all imagination in how care can be made *actually caring* dies.

In fact what social care at present is lacking is the very presence of *any* imagination that reaches toward the overarching and grandiose. For the overarching – even where it becomes overreaching - and the grandiose – especially when it veers toward the grandiloquent – are what have been key elements that have underpinned all the vast and meaningful advances in the creation of any socially based society – that is one in which the state takes an intimate interest, concern and action that seeks to improve on just the basic subsistence of existence within it, into one which actually creates and enhances lives that become worth the living.

For what was the idea of a creation of universal healthcare system free at the point of delivery if not overreaching. What was the creation of the welfare state if it was not grandiloquent? – The architects of both deployed vast imagination, incredible aims and the solid belief that society could be made better, not some time in future but right there and right then; that the future was, and is, made in the present, and delay is only the last refuge of the either the unworthy to govern or the bolt hole of the cravenly socially unconcerned.

Social care in the UK has for too long suffered from leaders of both varieties – the unworthy and the unconcerned – it is time for that to change, it is a change we should demand; and I would go further - it is a change that is that it is our *duty to compel* for if we do not we betray the legacy of imagination and will that created the society that has been ours to enjoy - and which we are so carelessly letting slip.

And so yes, some of the brief conclusions below are seemingly impossible, but they are not unimaginable, and – yes – they are within our reach if not quite within our grasp.

So the following will begin at the most forward thinking and vast and range down to smaller changes that I suggest would create a social care sector that does not habitually fail those who come to rely upon it.

1) <u>Nationalise All Social Care Provision.</u>

As the above will hopefully have made plain, social care, and most particularly care for the elderly – as this is where the fault lines in the private provision of care are most egregious – cannot be a profit seeking venture. The central conflict of interest between providing the best of care and the maximisation of profit is too great to bridge. As a guy is once supposed to have said, you cannot serve two masters, one will inevitably become the slave of the other - and in a business enterprise it is obvious which of these two aims is the slave and which is the master.

Some will argue that the market is better at providing "value for money" and "more efficient" than the public sector, but in social care what is "value for money" what is "efficiency"?

Less optimal care?

Rationing of services that should be seen as essential?

Employing carers with no thought to their suitability or aptitude?

Staffing levels so low that they stretch those same often unsuitable carers beyond their capacities?

All these points I have made as being present and at large in the private care sector and they are central to why care is now failing so many people.

The returning of all social care to the public sector is the fountainhead from which standards and quality of care can be driven up, all the following conclusions are easily within reach if we once more truly socialise care for the needy and vulnerable.

Many will say this aim would be impossible even in the best of times, never mind the straightened financial ones we find ourselves. And yet once more I refer back to the creation of the NHS and the modern welfare state – these were both created in the immediate post-war years when times were much harsher, the excuses for delay greater. Crises, as business books and publications are fond of lecturing us, also produce and open up possibilities, they create *opportunities* that more salad days do not present. We are at such a point now – we have the chance to *remake social care* from the ground up.

The mechanics of nationalising I leave, as it would create too much of a digression, but what I will posit here is that the for-profit provision of care is at present - and into the future - unsustainable, so the chances of bringing back social care into the public sector have never been more realistic. Business knows a good deal when it sees it and the chance to exit from a commercial activity that will become ever less rewarding financially will be only too gladly seized, the door at which we push against is unlocked at-to and only a little effort will open it right up.

2) <u>Create A Professional Representative and Registration Body for Care Workers.</u>

Throughout my story I have lamented so many times the lack of a professional body that registers, represents and holds to account Care Workers that this conclusion should be familiar and, hopefully, glaringly obvious.

How is it that we have reached a position where key care deliverers to our needy and vulnerable are unlicensed and unrepresented? If we demand that – say – Heavy Goods Vehicle (HGV) drivers must be properly trained and licensed – then why are those charged with the equally – if not more – responsible job of carrying out the most intimate and essential care tasks for those unable to do so for themselves - are not?

What does it say about our society that we take more of an interest in drivers of large vehicles than we do of those caring for the most vulnerable in our midst?

Further why is it that CW's who work alongside doctors and nurses in the health and social care environment are the only ones out of those professions that are not represented by professional membership body?

This makes no sense.

The Kingsmill Review suggests that CW's should be registered with the existing Healthcare Professions Council – HCPC- this would be a start; at least it would put CW's on some kind of equal footing with other healthcare workers and would provide some safeguards to ensure the quality of CW's through making registration mandatory for a licence to operate as a CW, as such it could prevent failing or abusive CW's from being allowed to carry on practicing.

However it does nothing as far as the equally important representational side of governing CW's. Therefore, as registration under the HCPC can and should implemented immediately, in the medium term a specific body should be set up for CW's along the lines of the General Medical Council – GMC – that exits its for doctors and the RCN for nurses.

This is vital because of the specific skillsets CW's are asked to possess and practice - and the particular strains and issues that apply to them as a class of healthcare workers - are manifestly different from other healthcare workers; therefore a specific body that reflects these demands, acknowledges these challenges and seeks to oversee, govern and – where necessary – defend CW's in the healthcare field is long overdue and should form a priority in the development and changes that the social care field is going to have to adapt to, one way or another, in the coming years.

3) **Pay All CW's The Living Wage.**

The private sector is able to boast a cost advantage – and therefore those imaginary "efficiency" and "value for money" gains - over public provision of care because it almost invariably pays CW's the minimum wage. This is a travesty of priorities as expanded upon in conclusion 1 above.

Care delivery - *at its point of delivery* - should not be administered by those whose value is held so low that they are paid the absolute lowest possible legal wage. Like it or not we live in a society that puts the value of work on a monetary basis and although most CW's do not do the job for financial advantage alone it is not a negligible consideration for any worker, let alone those doing an often very stressful and difficult job.

Low wages also encourage CW's – who are predominantly drawn from preexisting backgrounds of financial stress - to work ever longer hours to meet their most basic needs and financial commitments – this creates distorted incentives – which will be expanded upon further in then next conclusion – instead of providing good or excellent care for a limited number of hours CW's are instead tacitly driven to providing poor or failing care over an extended period.

The only way to counteract these distorted incentives is to pay those who actually deliver care a wage sufficient to meet their needs financially and so be able to concentrate better on meeting the needs of those they support or care for – the driver for care should be quality over quantity and this can only be achieved by in turn supporting CW's through enhanced wages.

A further benefit of better rewarding CW's reasonably would be that it would attract better and younger candidates for care work.

Both the Cavendish and Kingsmill Reviews focussed on the ageing nature of the care workforce and the difficulty it has of drawing younger applicants into care, although they focussed on the lack of any clear career progression as the main issue contributing to this factor and although they also both acknowledged that low wages were also a disincentive, they placed this below the fact that care work seemed to lead to nowhere professionally but more care work; I would – if I can make so bold – disagree about this ordering, I personally, and through experience, believe that making care work more rewarding remuneratively would attract better and younger candidates more effectively than some distant potential career path.

4) **Limit The Number Of Hours/Consecutive Days That Carers Work.**

As noted above, low wages create distorted incentives, the worst being that it encourages CW's to work ever more hours. I can personally bare testament through my own experience and my experiences with other CW's that performance declines exponentially after around 40hrs – 50hrs worked in any 7 day period.

Personally my care after working 3 twelve hour shifts on consecutive days diminished rapidly, and I can bear witness to the fact that this is almost an immutable law in care, any CW working above 50hrs in any 7 day period invariably did their job less effectively than than someone working less hours. Therefore those who worked sometimes 80 or 90hrs in a week consistently produced poor and negligent care – in short they did a lot of care hours badly than doing fewer care hours well.

Hours are not the only problem though, the number of days worked consecutively also adversely affects care performance. I firmly believe that because of the pressures and demands of care work - as well as the mental and emotional commitment that good care necessitates - that adequate time away from the care environment is key to keeping CW's focussed on providing good care. Anyone working more than 5 days in row, even if this is only for a few hours each day, also meet a point where their care capacities diminish vastly.

So it is that I believe that the number of hours/days any CW can work in any 7 day period should be capped. Personally I believe that 50-60hrs or 6 days consecutively should automatically trigger a two day rest period. This combined with conclusion 3 made above would greatly improve both CW performance and the quality of care that they deliver. We need to move away from the current system that values hours worked ahead of quality provided if we are to make any headway in creating a social care system that is professional, well organised and well run.

5) Unify Healthcare With Social Care To Create A Seamless Care System.

The gap between healthcare and social care that is both costly and damaging to the continuity of care for those who need it has already been recognised by politicians, so have the needless replication of processes and evaluations that draw so many resources away from actual caring. In the management argot there are "synergies" that could reap cost savings to made by closing the lacuna that exists currently which separates healthcare from social care.

Steps toward this goal have already been taken – some elements of social and healthcare care budgets have been folded together - however the changes are too piecemeal and the funding for initial costs of setting up a unified system too low to actually put what is currently an aspiration into reality.

This is incredibly shortsighted.

As noted above there are potentially huge saving to made by creating a seamless care system, but in order to reap them upfront costs need to be borne. If any government is serious about creating such a unified system they must realise they cannot do either on the fly or on the cheap. A well-funded transition will create a more robust and therefore enduring system, this should be the long term goal.

However cost savings are only one part of the equation, more important is the effect that having a unified care system will have, intimately, on its users. By bringing together health with social care far fewer people will have to suffer needlessly as they slip between the gap that currently exists. This should our ultimate endpoint – that care, however it is delivered, will not be a matter of organisations existing in separate "silos" but one that is as continuous as it is whole.

End.